Springer Series in Reliability Engineering

Series Editor

Professor Hoang Pham
Department of Industrial and Systems Engineering
Rutgers, The State University of New Jersey
96 Frelinghuysen Road
Piscataway, NJ 08854-8018
USA

Other titles in this series

The Universal Generating Function in Reliability Analysis and Optimization
Gregory Levitin

Warranty Management and Product Manufacture
D.N.P Murthy and Wallace R. Blischke

Maintenance Theory of Reliability
Toshio Nakagawa

System Software Reliability
Hoang Pham

Reliability and Optimal Maintenance
Hongzhou Wang and Hoang Pham

Applied Reliability and Quality
B.S. Dhillon

Shock and Damage Models in Reliability Theory
Toshio Nakagawa

Risk Management
Terje Aven and Jan Erik Vinnem

Satisfying Safety Goals by Probabilistic Risk Assessment
Hiromitsu Kumamoto

Offshore Risk Assessment (2nd Edition)
Jan Erik Vinnem

The Maintenance Management Framework
Adolfo Crespo Márquez

Human Reliability and Error in Transportation Systems
B.S. Dhillon

Complex System Maintenance Handbook
D.N.P. Murthy and Khairy A.H. Kobbacy

Recent Advances in Reliability and Quality in Design
Hoang Pham

Product Reliability
D.N. Prabhakar Murthy, Marvin Rausand and Trond Østerås

B.S. Dhillon

Mining Equipment Reliability, Maintainability, and Safety

Springer

B.S. Dhillon, PhD
Department of Mechanical Engineering
University of Ottawa
Ottawa
Ontario K1N 6N5
Canada

ISBN 978-1-84800-287-6 e-ISBN 978-1-84800-288-3

DOI 10.1007/978-1-84800-288-3

Springer Series in Reliability Engineering ISSN 1614-7839

British Library Cataloguing in Publication Data
Dhillon, B. S. (Balbir S.), 1947 –
Mining equipment reliability, maintainability, and safety.
– (Springer series in reliability engineering)
1. Mining machinery – Maintainability 2. Mining machinery –
Reliability 3. Mining machinery – Safety measures
I. Title
622'.028
ISBN-13: 9781848002876

Library of Congress Control Number: 2008929518

Cover design: deblik, Berlin, Germany

Printed on acid-free paper

9 8 7 6 5 4 3 2 1

springer.com

This book is affectionately dedicated to my dear friend Dr. S.N. Rayapati for his stimulating conversations, support, and friendship over the years.

Preface

The history of mining may be traced back to the ancient Egyptians, who operated malachite mines. Today a large number of people are employed in the mining industry throughout the world. For example, in the USA alone around 675,000 people work in the natural resources and mining sector. Today, the mining industry uses various types of complex and sophisticated equipment whose reliability, maintainability, and safety have become an important issue.

Although over the years a large number of journal and conference proceedings articles on mining equipment reliability, maintainability, and safety have appeared, to the best of author's knowledge, there is no book that covers all three of these topics within its framework. This causes a great deal of difficulty for information seekers on the subjects because they must consult many different and diverse sources.

Thus, the main objective of this book is to combine all three of these topics into a single volume, to eliminate the need to consult many different and diverse sources in obtaining desired information. The sources of most of the material presented are given in the reference section at the end of each chapter. This will be useful to readers if they desire to delve deeper into a particular area. The book contains a chapter on mathematical concepts and another chapter on introductory material on reliability, maintainability, and safety considered essential to understand contents of subsequent chapters.

The topics covered in the volume are treated in such a manner that the reader will require no previous knowledge to understand the contents. At appropriate places, the book contains examples along with their solutions, and at the end of each chapter there are numerous problems to test reader comprehension.

An extensive list of references on mining equipment reliability, maintainability, and safety is provided at the end of the book to give readers a view of developments in the area over the years.

The book is composed of 11 chapters. Chapter 1 presents the need for improving mining equipment reliability, maintainability, and safety; mining-equipment-related facts and figures, important terms and definitions, and useful information on mining equipment reliability, maintainability, and safety classified under six distinct categories. Chapter 2 is devoted to mathematical concepts considered useful for performing mining equipment reliability, maintainability, and safety analysis. It covers topics such as Boolean algebra laws, probability properties, useful mathematical definitions, and probability distributions.

Chapter 3 presents various introductory aspects of reliability, maintainability, and safety including reliability networks, commonly used methods in reliability analysis, maintainability functions, maintainability analysis tools, safety analysis methods, and safety indexes. Chapter 4 is devoted to mining equipment reliability and covers topics such as reasons for improving mining equipment reliability, open-pit system reliability analysis, programmable electronic mining system failures, fault tree analysis of shovel machine, and dump-truck tire reliability and factors affecting their reliability. Various aspects of human factors and error in mining are covered in Chap. 5. Some of the topics covered in the chapter are the need for human-factor application in mining, human sensory capacities, human-factor formulas, useful general human-factor guidelines for application in mining equipment design, classifications and causes of human errors leading to fatal accidents in mines, typical mining equipment maintenance errors, useful design improvements to reduce mining equipment maintenance errors, and human error analysis methods for application in the area of mining.

Chapters 6 and 7 are devoted to mining equipment maintainability and mining equipment reliability and maintainability testing, respectively. Chapter 6 covers topics such as reliability test classifications, success testing, accelerated testing, confidence interval estimates for mining equipment mean time between failures, test methods to obtain maintainability-related test data for mining equipment, test methods for demonstrating diverse maintainability parameters, and useful guidelines for avoiding pitfalls in maintainability testing of mining equipment. Some of the topics covered in Chap. 7 are the meanings of the mining equipment maintainability and design-induced maintainability problems of mining equipment, advantages of the improved mining equipment maintainability design, mining equipment maintainability design characteristics, maintainability measures for mining equipment, and common maintainability design errors and useful maintainability design guideline for mining equipment.

Chapter 8 presents various important aspects of mining equipment maintenance including maintenance-related facts and figures, factors contributing to equipment maintenance cost in mines, maintenance of explosion-protected switchgear in mines, useful maintenance measures for mines, and mathematical models for performing mining equipment maintenance. Various important aspects of mining equipment costing are covered in Chap. 9. Some of the topics covered include reasons for mining equipment costing, methods for making mining equipment investment decisions, cost estimation models for mining equipment, life cycle costing concept, and life cycle cost estimation models for mining equipment.

Chapter 10 is devoted to the introductory aspects of mining equipment safety and covers topics such as facts and figures, quarry accidents, causes of mining equipment accidents and major sources of mining equipment fires, methods for performing mining equipment safety analysis, human-factor-related tips for safer mining equipment, strategies to reduce mining equipment fires and injuries, and general areas for safety improvements in mines. Finally, Chap. 11 presents various important aspects of programmable electronic mining system safety including programmable-electronic-related mishaps, methods for performing hazard and risk analysis of programmable electronic mining systems, lessons learned in addressing the safety of programmable electronic mining systems, and sources for obtaining programmable electronic mining system safety-related information.

This book will be useful to many individuals including engineering professionals working in the mining industry, mining administrators, mining engineering undergraduate and graduate students, mining engineering researchers and instructors, reliability, maintainability, maintenance, human factors, and safety professionals, and design engineers and associated professionals concerned with mining equipment.

The author is deeply indebted to many colleagues and students for their interest throughout this project. The invisible contributions of my children, Jasmine and Mark, are also appreciated. Last but not least, I thank my other half, friend, and wife, Rosy, for typing various portions of the book and other related materials, and for her timely help in proofreading and tolerance.

Ottawa, Ontario *B.S. Dhillon*

Contents

Chapter 1
Introduction

1.1 The Need for Improving Mining Equipment Reliability, Maintainability, and Safety

The history of mining may be traced back to the ancient Egyptians, who operated malachite mines at Wady Maghareh on the Sinai Peninsula and at Timna in the Negev. Today millions of people are employed in the mining industry throughout the world. For example, in the USA alone around 675,000 people are employed in the natural resources and mining sector [1].

Each year billions of dollars are spent to produce various types of equipment for use by the mining industry throughout the world, and this expenditure is increasing rapidly. For example, in 2004 American mining equipment manufacturers shipped around $1.4 billion worth of goods and a year later, in 2005, the figure jumped to $2 billion [2]. Nowadays, the competitive global economy is forcing mining companies around the world to modernize its operations through increased mechanization and automation.

Thus, mining equipment is becoming more complex and sophisticated, and its cost is increasing at an alarming rate. In addition, its safety-related issues are receiving increased attention. This in turn makes it cost ineffective to have backup units nor unsafe units. All in all, in order to meet production targets and tight schedules, the mining companies are increasingly demanding better equipment reliability, maintainability, and safety.

1.2 Mining-equipment-related Facts and Figures

This section presents facts and figures directly or indirectly concerned with mining equipment reliability, maintainability, and safety.

- In 2006, the value of shipments of selected types of mining machinery and related equipment from American manufacturers was estimated to be around $2.5 billion [3].
- The average fatality rate in the US mining industry was 27.7 per 100,000 workers, as opposed to 4.8 per 100,000 workers in all US mining industry sectors during the period 1992–2001.
- Equipment maintenance costs range from 20% to over 35% of total mine operating costs [5].
- Approximately 10% of production time is lost by unplanned maintenance in the Australian underground coal mining industry [6].
- Annual mining deaths in the USA decreased from around 500 in the late 1950s to about 93 during the 1990s [7].
- In 2004, around 17% of the 37,445 injuries in American underground coal mines were associated with bolting machines [8].
- During the period 1995–2001, a total of 11 programmable electronic-related mining incidents were reported in the USA; 4 of these were fatalities [9, 10].
- In open-pit mines in both Chile and Indonesia, maintenance cost is more than 60% of the operating cost [11].
- According to various civilian and military studies, it is possible to reduce preventive maintenance and corrective maintenance task times by 40% to 70% with planned maintainability design efforts [5].
- During the period 1990–1999, electricity was the fourth leading cause of death in the US mining industry [12].
- Usually, the cost of maintenance in the mining industry varies from 40 to 50% of the equipment operating cost [13].
- During the period 1978–1988, maintenance activity accounted for 34% of all lost time injuries in US mines [14].
- During the period 1983–1990, with respect to per-injury data for independent contractor employees in the mining industry, approximately 20% of the coal-mine-related injuries occurred during machine-maintenance activity or while using hand tools [15].
- During the period 1990–1999, 197 equipment fires caused 76 injuries in coal mining operations in the USA [16].

1.3 Terms and Definitions

There are a large number of terms and definitions used in the area of reliability, maintainability, and safety. This section presents some of the terms and definitions considered useful for application in the area of mining equipment reliability, maintainability, and safety [17–21].

- **Accident:** an unplanned and undesired act
- **Availability:** the probability that a piece of equipment/system is functioning satisfactorily at time t when used according to specified conditions, where the total

time includes operating time, logistical time, active repair time, and administrative time

- **Failure:** the inability of an item/piece of equipment/system to operate within specified guidelines
- **Failure mode:** the abnormality of item/equipment/system performance that causes the item/piece of equipment/system to be considered as having failed
- **Hazard:** the source of energy and the behavioral and physiological factors that, when not controlled effectively, lead to harmful incidents
- **Hazard rate:** the ratio of the change in the number of items that have malfunctioned to the number of items that have survived at time t
- **Maintainability:** the probability that a failed system/piece of equipment/item will be restored to its satisfactory operating state
- **Maintenance:** all actions necessary to retain an item/piece of equipment/system in, or restore it to, a specified condition
- **Mean time to failure (exponential distribution):** the sum of the operating time of given items divided by the total number of failures
- **Mean time to repair:** a figure of merit depending on item/equipment/system maintainability equal to the mean item/equipment/system repair time; in the case of exponentially distributed times to repair, mean time to repair is the reciprocal of the repair rate
- **Mine:** an excavation from which minerals or ore is extracted
- **Mission time:** the time during which an item/piece of equipment/system is performing its specified function
- **Open-pit mining:** a form of operation designed for extracting minerals that lie near the Earth's surface
- **Ore:** any natural combination of minerals
- **Redundancy:** the existence of more than one means to accomplish a specified function
- **Reliability:** the probability that an item/piece of equipment/system will carry out its specified mission satisfactorily for the stated time period when used under specified conditions
- **Reliability model:** a model used to assess, estimate, or predict reliability
- **Safety:** the conservation of human life and its effectiveness and the prevention of damage to items/equipment/systems as per stated requirements
- **Safety assessment:** the quantitative/qualitative determination of safety
- **Safety process:** a set of procedures followed to enable the safety requirements of an item/piece of equipment/system to be identified and met
- **Unsafe condition:** any condition, under the right circumstances, that will lead to an accident

1.4 Useful Information on Mining Equipment Reliability, Maintainability, and Safety

There are many different sources for obtaining mining equipment reliability-, maintainability-, and safety-related information. This section lists some of the most useful sources, directly or indirectly, for obtaining such information, under a number of distinct categories.

1.4.1 Journals and Magazines

- Journal of Quality in Maintenance Engineering
- Reliability Engineering and System Safety
- IEEE Transactions on Reliability
- International Journal of Reliability, Quality, and Safety Engineering
- Professional Safety
- Hazard Prevention
- Accident Analysis and Prevention
- Reliability Review
- Journal of Fire Safety
- Reliability: The Magazine for Improved Plant Reliability
- Quality and Reliability Engineering International
- RAMS ASIA (Reliability, Availability, Maintainability, and Safety (RAMS) Quarterly Journal)
- National Safety News
- Engineering and Mining Journal
- Australian Mining
- Mining Magazine
- Engineers Australia
- Mining Technology
- Soviet Mining Science
- International Journal of Surface Mining & Reclamation
- Mining Congress Journal
- Coal Age
- Transactions of the Canadian Institute of Mining and Metallurgy
- CIM Bulletin
- The Mining Engineer
- Engineering Failure Analysis

1.4.2 Conference Proceedings

- Proceedings of the Annual Institute on Mining Health, Safety, and Research
- Proceedings of the Annual Reliability and Maintainability Symposium
- Proceedings of the ISSAT International Conferences on Reliability and Quality in Design
- Proceedings of the European Conferences on Safety and Reliability
- Proceedings of the International Conferences on Probabilistic Safety Assessment and Management.
- Proceedings of the International Conference on Reliability, Production, and Control in Coal Mines, 1991
- Proceedings of the Annual Meetings of the Society for Mining, Metallurgy, and Exploration
- Proceedings of the International Symposium on Mine Planning and Equipment Selection
- Proceedings of the American Mining Congress-Coal Convention, 1991
- Proceedings of the IEEE Annual Industry Applications Conferences

1.4.3 Books

- Blanchard, B.S., Verma, D., Peterson, E.L.: Maintainability: A Key to Effective Serviceability and Maintenance Management. Wiley, New York (1995)
- Cox, S.J.: Reliability, Safety, and Risk Management: An Integrated Approach. Butterworth-Heinemann, New York (1991)
- Dhillon, B.S.: Design Reliability: Fundamentals and Applications. CRC Press, Boca Raton, FL (1999)
- Dhillon, B.S.: Engineering Maintainability. Gulf Publishing, Houston, TX (1999)
- Dhillon, B.S.: Engineering Safety: Fundamentals, Techniques, and Applications. World Scientific, River Edge, NJ (2003)
- Goetsch, D.L.: Occupational Safety and Health. Prentice-Hall, Englewood Cliffs, NJ (1996)
- Grant Ireson, W., Coombs, C.F., Moss, R.Y. (eds.): Handbook of Reliability Engineering and Management. McGraw-Hill, New York (1996)
- Hammer, W., Price, D.: Occupational Safety and Engineering. Prentice-Hall, Upper Saddle River, NJ (2001)
- Moubray, J.: Reliability-Centered Maintenance. Industrial, New York (1997)
- Shooman, M.L.: Probabilistic Reliability: An Engineering Approach. McGraw-Hill, New York (1968)

1.4.4 Organizations

- American Society of Safety Engineers, 1800 East Oakton St., Des Plaines, IL, USA
- World Safety Organization, P.O. Box No. 1, Lalong Laan Building, Pasay City, Metro Manila, The Philippines
- Reliability Society, IEEE, P.O. Box 1331, Piscataway, NJ, USA
- The American Institute of Mining, Metallurgical, and Petroleum Engineers, 8307 Shaffer Parkway, Littleton, CO, USA
- Society for Maintenance and Reliability Professionals, 401 N. Michigan Avenue, Chicago, IL, USA
- Society of Logistics Engineers, 8100 Professional Place, Suite 211, Hyattsville, MD, USA
- Society for Machinery Failure Prevention Technology, 4193 Sudley Road, Haymarket, VA, USA
- National Safety Council, 444 North Michigan Avenue, Chicago, IL, U.S.A.
- American Society for Quality, Reliability Division, 600 North Plankinton Avenue, Milwaukee, WI, USA
- System Safety Society, 14252 Culver Drive, Suite A-261, Irvine, CA, USA

1.4.5 Standards

- MIL-STD-785, Reliability Program for Systems and Equipment, Development and Production, US Department of Defense, Washington, DC
- MIL-STD-1629, Procedures for Performing Failure Mode, Effects and Criticality Analysis, US Department of Defense, Washington, DC
- MIL-HDBK-217, Reliability Prediction of Electronic Equipment, US Department of Defense, Washington, DC
- MIL-HDBK-781, Reliability Test Methods, Plans, and Environments for Engineering Development, Qualification, and Production, US Department of Defense, Washington, DC
- MIL-STD-882, Systems Safety Program for System and Associated Sub-system and Equipment-Requirements, US Department of Defense, Washington, DC
- IEC 60950, Safety of Information Technology Equipment, International Electro-Technical Commission (IEC), Geneva
- MIL-STD-470, Maintainability Program for Systems and Equipment, US Department of Defense, Washington, DC
- MIL-HDBK-472, Maintainability Prediction, US Department of Defense, Washington, DC
- MIL-STD-471, Maintainability/Verification/Demonstration/Evaluation, US Department of Defense, Washington, DC

- IEC 61508 SET, Functional Safety of Electrical/Electronic/Programmable Electronic Safety-related Systems Parts 1–7, International Electrotechnical Commission (IEC), Geneva

1.4.6 Data Information Sources

- RAC EEMD1, Electronic Equipment Maintainability Data, Reliability Analysis Center, Rome Air Development Center, Griffiss Air Force Base, Rome, NY, USA
- Government Industry Data Exchange Program (GIDEP), GIDEP Operations Center, US Department of the Navy, Corona, CA, USA
- National Technical Information Service (NTIS), 5285 Port Royal Road, Springfield, VA, USA
- Safety Research Information Service, National Safety Council, 444 North Michigan Avenue, Chicago, IL, USA
- Defense Technical Information Center, DTIC-FDAC, 8725 John J. Kingman Road, Suite 0944, Fort Belvoir, VA, USA
- National Aeronautics and Space Administration (NASA) Parts Reliability Information Center, George C. Marshall Space Flight Center, Huntsville, AL, USA

1.5 Problems

1. Discuss the need for improving mining equipment reliability, maintainability, and safety.
2. List at least seven facts and figures concerned with mining equipment reliability, maintainability, and safety.
3. Define the following terms:
   - Open-pit mining
   - Hazard rate
   - Reliability
4. List the five most important journals that publish mining-equipment-reliability-related studies.
5. Compare mining equipment reliability with its maintainability.
6. List at least four sources that can be useful for obtaining mining equipment reliability-, maintainability-, and safety-related information.
7. List at least four of the most important standards on reliability, maintainability, and safety.
8. Define the following terms:
   - Safety
   - Failure
   - Accident

9. What is mean time to repair?
10. Compare equipment availability with equipment reliability.

References

1. Division of Information Services, United States Bureau of Labor Statistics, 2 Massachusetts Avenue, Washington, DC
2. Chadwick, J., Higgins, S.: US technology. Int. Min. September, 44–54 (2006)
3. Manufacturing and Construction Division, US Census Bureau, US Department of Commerce, Washington, DC
4. Iannacchione, A., Mucho, T.: 100 years of improvement in aggregate worker safety. Stone Sand Gravel Rev. pp. 28–34 (March/April 2003)
5. Unger, R.L., Conway, K.: Impact of maintainability design on injury rates and maintenance costs for underground mining equipment. In: Improving Safety at Small Underground Mines, compiled by R.H. Peters, Special Publication Report No. 18–94, US Bureau of Mines, Washington, DC, pp. 140–167 (1994)
6. Clark, D.: Tribology – its applications to equipment reliability and maintainability design in the underground coal mining industry. In: Proceedings of the Institution of Engineers Australia Tribology Conference, pp. 38–44 (1990)
7. Historical Data on Mine Disasters in the United States, US Department of Labor, Washington, DC. Available online at www.msha.gov/msainfo/factsheets/mshafacts8.htm
8. Burgess-Limerick, R., Steiner, L.: Preventing injuries: analysis of injuries highlights high priority hazards associated with underground coal mining equipment. Am. Longwall Mag., pp. 19–20 (August 2006)
9. Sammarco, J.J.: Addressing the safety of programmable electronic mining systems: lessons learned. In: Proceedings of the IEEE 37th Industry Applications Society Meeting, pp. 692–698 (2003)
10. Fatal Alert Bulletins, Fatal Grams and Fatal Investigation Reports, Mine Safety and Health Administration (MSHA), Washington, DC (May 2001). Available online at www.msha.gov/fatals/fab.htm
11. Hall, R.A., Daneshmend, L.K., Lipsett, M.G., Wong, J.: Reliability analysis as a tool for surface mining equipment evaluation and selection. CIM Bull. 93, 1044, 78–82 (2000)
12. Cawley, J.C.: Electrical accidents in the mining industry, 1990–1999. IEEE Trans. Ind. Appl. 39, 6, 1570–1576 (2003)
13. Forsmann, B., Kumar, U.: Surface mining equipment and maintenance trends in the scandinavian countries. J. Mines Minerals Fuels August/September 266–269 (1992)
14. MSHA Data for 1978–1988, Mine Safety and Health Administration (MSHA), US Department of Labor, Washington, DC
15. Rethi, L.L., Barett, E.A.: A summary of injury data for independent contractor employees in the mining industry from 1983–1990. Report No. USBMIC 9344, US Bureau of Mines, Washington, DC (1983)
16. De Rosa, M.: Equipment fires cause injuries: recent NIOSH study reveals trends for equipment fires at U.S. coal mines. Coal Age pp. 28–31 (October 2004)
17. MIL-STD-721, Definitions of Effectiveness Terms for Reliability, Maintainability, Human Factors, and Safety, US Department of Defense, Washington, DC
18. Omdahl, T.P. (ed.): Reliability, Availability, and Maintainability (RAM) Dictionary. ASQC Quality Press, Milwaukee, WI (1988)
19. Naresky, J.J.: Reliability Definitions. IEEE Trans. Reliabil. 19, 198–200 (1970)

20. Dictionary of Terms Used in the Safety Profession, 3rd ed., American Society of Safety Engineers, Des Plaines, IL (1988)
21. McKenna, T., Oliverson, R.: Glossary of Reliability and Maintenance Terms. Gulf Publishing, Houston, TX (1997)

Chapter 2
Introductory Mathematical Concepts for Mining Equipment Reliability, Maintainability, and Safety Analysis

2.1 Introduction

As in other areas of engineering analysis, various mathematical concepts play a pivotal role in mining equipment reliability, maintainability, and safety analysis. Although the history of our current number symbols can be traced back to the stone columns erected by the Scythian emperor Asoka of India in 250 B.C., the application of mathematical concepts in engineering in general is relatively new [1].

In particular, probability plays a central role in the analysis of mining equipment reliability, maintainability, and safety problems; its history may only be traced back to the 16th-century writings of Girolamo Cardano (1501–1576) [1, 2]. In these writings, Cardano considered some interesting questions on probability. In the 17th century, the problem of dividing the winnings in a game of chance was solved independently and correctly by Blaise Pascal (1623–1662) and Pierre Fermat (1601–1665). In the 18th century, probability concepts were further developed and successfully applied to areas other than games of chance by Pierre Laplace (1749–1827) and Karl Gauss (1777–1855) [2, 3].

A detailed history of mathematics including probability is available in Refs. [1, 2]. This chapter presents various introductory mathematical concepts considered useful for performing mining equipment reliability, maintainability, and safety analysis [4, 5].

2.2 Range, Arithmetic Mean, Mean Deviation, and Standard Deviation

Many statistical measures are used to analyze reliability-, maintainability-, and safety-related data. This section presents a number of such measures considered useful for application in the area of mining equipment reliability, maintainability, and safety.

2.2.1 Range

This is a measure of dispersion or variation. More specifically, the range of a data set is the difference between the largest and the smallest values in the set.

Example 2.1

A mining facility reported the following monthly equipment failures over a period of 12 months:

$$40,\ 5,\ 10,\ 15,\ 20,\ 46,\ 50,\ 19,\ 25,\ 17,\ 35,\ \text{and } 16\,.$$

Find the range of the above data set values.

By examining the given data values, we conclude that the largest and the smallest values are 50 and 5, respectively. Thus, the range, R, of the given data set is expressed by

$$R = \text{Largest value} - \text{Smallest value} = 50 - 5 = 45\,.$$

Thus, the range of the given data set is 45.

2.2.2 Arithmetic Mean

The arithmetic mean is defined by

$$m = \frac{\sum_{j=1}^{n} m_j}{n}\,, \tag{2.1}$$

where

m is the mean value,
m_j is the data value j; for $j = 1, 2, 3, \ldots, n$,
n is the total number of data values.

Example 2.2

A mining equipment manufacturing organization inspected ten identical mining systems and found 5, 10, 3, 2, 7, 15, 20, 1, 9, and 8 defects in each system. Calculate the average number of defects per mining system (*i.e.*, arithmetic mean of the data set).

Inserting the specified data values into Eq. (2.1) we obtain

$$m = \frac{5+10+3+2+7+15+20+1+9+8}{10} = 8\,.$$

Thus, the average number of defects per mining system (*i.e.*, arithmetic mean of the data set) is 8.

2.2.3 Mean Deviation

This is a widely used measure of dispersion that indicates the degree to which given data tend to spread about a mean value. The mean deviation is defined by

$$m_{\mathrm{d}} = \frac{\sum_{j=1}^{n} \left|m_j - m\right|}{n}\,, \tag{2.2}$$

where

n	is the total number of data values,
m_j	is the data value j; for $j = 1, 2, 3, \ldots, n$,
m_d	is the mean deviation,
m	is the mean value,
$\left\|m_j - m\right\|$	is the absolute value of the deviation of m_j from m.

Example 2.3

Calculate the mean deviation of the data values given in Example 2.2. Using the given data and calculated values of Example 2.2 in Eq. (2.2) yields

$$\begin{aligned} m_{\mathrm{d}} &= \frac{|5-8|+|10-8|+|3-8|+|2-8|+|7-8|+|15-8|}{10} \\ &\quad + \frac{|20-8|+|1-8|+|9-8|+|8-8|}{10} \\ &= \frac{3+2+5+6+1+7+12+7+1+0}{10} \\ &= 4.4\,. \end{aligned}$$

Thus, the mean deviation of the given data values is 4.4.

2.2.4 Standard Deviation

This is another measure of dispersion of data in a data set about the mean value. The standard deviation is defined by [3]

$$\sigma = \left[\frac{\sum_{j=1}^{n} (m_j - m)^2}{n} \right]^{1/2}, \tag{2.3}$$

where

σ is the standard deviation.

The following three standard deviation properties are associated with the normal distribution presented later in the chapter:

- 99.73% of the all data values are included between $m - 3\sigma$ and $m + 3\sigma$.
- 95.45% of the all data values are included between $m - 2\sigma$ and $m + 2\sigma$.
- 68.27% of the all data values are included between $m - \sigma$ and $m + \sigma$.

Example 2.4

Calculate the standard deviation of the data values given in Example 2.2.

Using the given data and calculated value of Example 2.2 in Eq. (2.3) we get

$$\sigma = \left[\frac{(5-8)^2 + (10-8)^2 + (3-8)^2 + (2-8)^2 + (7-8)^2 + (15-8)^2}{10} \right.$$

$$\left. + \frac{(20-8)^2 + (1-8)^2 + (9-8)^2 + (8-8)^2}{10} \right]^{1/2}$$

$$= \left[\frac{9+4+25+36+1+49+144+49+1+0}{10} \right]^{1/2} = 5.64\,.$$

Thus, the standard deviation of the data values given in Example 2.2 is 5.64.

2.3 Boolean Algebra Laws and Probability Definition and Properties

Boolean algebra plays an important role in probability theory and is named after mathematician George Boole (1813–1864). Some of its laws are as follows [6, 7]:

$$C + D = D + C\,, \tag{2.4}$$

where

C is a set or an event,
D is a set or an event,
$+$ denotes the union of events or sets

$$C \cdot D = D \cdot C \,, \tag{2.5}$$

where

dot between C and D or D and C denotes the intersection of events or sets. Sometimes the intersection of events is written without the dot (*e.g.*, CD), but it still conveys exactly the same meaning.

$$DD = D \,, \tag{2.6}$$

$$C + C = C \,, \tag{2.7}$$

$$C(C + D) = C \,, \tag{2.8}$$

$$D + DC = D \,, \tag{2.9}$$

$$C(D + E) = CD + CE \,, \tag{2.10}$$

where

E is a set or an event.

$$C + 0 = C \,, \tag{2.11}$$

$$(C + D)(C + E) = C + DE \,. \tag{2.12}$$

Probability may be defined as the likelihood of occurrence of a given event. Mathematically, it is expressed as follows [8]:

$$P(Y) = \lim_{n \to \infty} \left[\frac{M}{n} \right] \,, \tag{2.13}$$

where

$P(Y)$ is the probability of occurrence of event Y,
M is the number of times event Y occurs in the n repeated experiments.

Some probability properties are as follows [8]:

- The probability of occurrence of event, say X, is

$$0 \leq P(X) \leq 1 \,. \tag{2.14}$$

- The probability of occurrence and nonoccurrence of an event, say X, is always

$$P(X) + P(\bar{X}) = 1 \,, \tag{2.15}$$

where

$P(X)$ is the probability of occurrence of event X,
$P(\bar{X})$ is the probability of nonoccurrence of event X.

- The probability of an intersection of K independent events is given by

$$P(X_1X_2X_3\ldots X_K) = P(X_1)P(X_2)P(X_3)\cdot\ldots\cdot P(X_K)\,, \tag{2.16}$$

where

$P(X_i)$ is the probability of occurrence of event X_i, for $i = 1, 2, 3, \ldots, K$.

- The probability of the union of K independent events is expressed by

$$P(X_1 + X_2 + \ldots + X_K) = 1 - \prod_{i=1}^{K}(1 - P(X_i))\,. \tag{2.17}$$

- The probability of the union of K mutually exclusive events is

$$P(X_1 + X_2 + \ldots + X_K) = \sum_{i=1}^{K} P(X_i)\,. \tag{2.18}$$

Example 2.5

Assume that in Eqs. (2.17) and (2.18) we have $K = 2$, $P(X_1) = 0.05$, and $P(X_2) = 0.12$. Calculate the probability of the union of events X_1 and X_2 using Eqs. (2.17) and (2.18) and comment on the resulting probability values.

Inserting the given data into Eq. (2.17) we get

$$\begin{aligned} P(X_1 + X_2) &= P(X_1) + P(X_2) - P(X_1)P(X_2) \\ &= 0.05 + 0.12 - (0.05)(0.12) \\ &= 0.164\,. \end{aligned}$$

Using the specified data values in Eq. (2.18) we get

$$\begin{aligned} P(X_1 + X_2) &= P(X_1) + P(X_2) \\ &= 0.05 + 0.12 \\ &= 0.17\,. \end{aligned}$$

This means that the probability of the union of mutually exclusive events X_1 and X_2 is higher than the probability of the union of independent events X_1 and X_2.

2.4 Useful Mathematical Definitions

This section presents a number of mathematical definitions considered useful to perform reliability, maintainability, and safety studies in the mining industry [3, 8].

2.4.1 Cumulative Distribution Function

For continuous random variables, the cumulative distribution function is defined by

$$F(t) = \int_{-\infty}^{t} f(x)\,\mathrm{d}x\,, \tag{2.19}$$

where

$F(t)$ is the cumulative distribution function,
$f(x)$ is the probability density function of continuous random variable x,
t is time.

For $t = \infty$, Eq. (2.19) yields

$$F(\infty) = \int_{-\infty}^{\infty} f(x)\,\mathrm{d}x = 1\,. \tag{2.20}$$

This simply means that the total area under the probability density curve is equal to unity.

Usually, in reliability work Eq. (2.19) is simply expressed as

$$F(t) = \int_{0}^{t} f(x)\,\mathrm{d}x\,. \tag{2.21}$$

2.4.2 Probability Density Function

For continuous random variables, the probability density function is defined by

$$f(t) = \frac{\mathrm{d}F(t)}{\mathrm{d}t}\,, \tag{2.22}$$

where

$f(t)$ is the probability density function (in reliability work, it is often called failure density function).

2.4.3 Reliability Function

The reliability function is defined by

$$R(t) = 1 - F(t) = 1 - \int_0^t f(x)\,\mathrm{d}x\,, \tag{2.23}$$

where

$R(t)$ is the reliability function or simply reliability at time t.

2.4.4 Expected Value

For continuous random variables, this is defined by

$$E(t) = m = \int_{-\infty}^{\infty} t f(t)\,\mathrm{d}t\,, \tag{2.24}$$

where

$E(t)$ is the expected value of the continuous random variable t.
m is the mean of the continuous random variable t. In reliability work, it is referred as mean time to failure.

2.4.5 Variance

The variance of a random variable t is defined by

$$\sigma^2(t) = E(t^2) - [E(t)]^2 \tag{2.25}$$

or

$$\sigma^2(t) = \int_0^{\infty} t^2 f(t)\,\mathrm{d}t - m^2\,, \tag{2.26}$$

where

$\sigma^2(t)$ is the variance of random variable t.

2.4.6 Laplace Transform

This is defined by

$$f(s) = \int_0^\infty f(t)\,\mathrm{e}^{-st}\,\mathrm{d}t\,, \tag{2.27}$$

where

s	is the Laplace transform variable,
t	is the time variable,
$f(s)$	is the Laplace transform of the function $f(t)$.

Laplace transforms of some commonly occurring functions in mining equipment reliability, maintainability, and safety studies are presented in Table 2.1 [9, 10].

2.4.7 Laplace Transform: Final Value Theorem

If the following limits exist, then the final-value theorem may be stated as

$$\lim_{t\to\infty} f(t) = \lim_{s\to 0}\,[sf(s)]\,. \tag{2.28}$$

Table 2.1 Laplace transforms of some commonly occurring functions in mining equipment reliability, maintainability, and safety studies

$f(t)$	$f(s)$
$\mathrm{e}^{-\lambda t}$	$\dfrac{1}{(s+\lambda)}$
$t\,\mathrm{e}^{-\lambda t}$	$\dfrac{1}{(s+\lambda)^2}$
$tf(t)$	$-\dfrac{\mathrm{d}f(s)}{\mathrm{d}s}$
c (a constant)	$\dfrac{c}{s}$
$\dfrac{\mathrm{d}f(t)}{\mathrm{d}t}$	$sf(s)-f(0)$
t^m, for $m = 1, 2, 3, \ldots$	$\dfrac{m!}{s^{m+1}}$
$\int_0^t f(t)\,\mathrm{d}t$	$\dfrac{f(s)}{s}$

2.5 Probability Distributions

Over the years, a large number of probability distributions have been developed to perform various types of statistical analysis [11]. This section presents some of these probability distributions considered useful for application in the area of mining equipment reliability, maintainability, and safety.

2.5.1 *Binomial Distribution*

This is a discrete random variable distribution and it was developed by Jakob Bernoulli (1654–1705) [1]. Thus, it is also called a Bernoulli distribution. The distribution is used in situations where one is concerned with the probabilities of outcome such as the total number of occurrences (*e.g.*, failures) in a sequence of given number of trials. More specifically, each of these trials has two possible outcomes (*e.g.*, success or failure), but the probability of each trial remains constant or unchanged.

The binomial probability density function, $f(x)$, is defined by

$$f(x) = \frac{m!}{x!(m-x)!} p^x q^{m-x} , \quad \text{for } x = 0, 1, 2, 3, \ldots, m , \tag{2.29}$$

where

p is the single trial probability of occurrence (*e.g.*, success),
q is the single trial probability of nonoccurrence (*e.g.*, failure),
x is the number of nonoccurrences (*e.g.*, failures) in m trials.

The cumulative distribution function is given by [8, 11]

$$F(x) = \sum_{i=0}^{x} \frac{m!}{i!(m-i)!} p^i q^{m-i} , \tag{2.30}$$

where

$F(x)$ is the probability of x or less nonoccurrences in m trials.

2.5.2 *Exponential Distribution*

This is a continuous random variable distribution and is widely used in reliability, maintainability, and safety work. Two principal reasons for its widespread use are as follows:

- Easy to handle in performing various types of reliability, maintainability, and safety analyses.

- Constant failure rates of many engineering items during their useful life periods, particularly electronic ones [12].

The distribution probability density function is expressed by

$$f(t) = \lambda\,\mathrm{e}^{-\lambda t}\,, \quad \text{for } \lambda > 0,\ t \geq 0\,, \tag{2.31}$$

where

$f(t)$	is the probability density function,
t	is time,
λ	is the distribution parameter. In reliability work, it is known as the constant failure rate.

Substituting Eq. (2.31) into Eq. (2.21) we get the following expression for the cumulative distribution function:

$$F(t) = \int_0^t \lambda\,\mathrm{e}^{-\lambda x}\,\mathrm{d}x = 1 - \mathrm{e}^{-\lambda t}\,. \tag{2.32}$$

2.5.3 Rayleigh Distribution

This continuous random variable distribution is named after its originator, John Rayleigh (1842–1919) [1]. The distribution probability density function is defined by

$$f(t) = \frac{2}{\alpha^2}t\,\mathrm{e}^{-\left(\frac{t}{\alpha}\right)^2}\,, \quad \text{for } \alpha > 0,\ t \geq 0\,, \tag{2.33}$$

where

α is the distribution parameter.

Using Eq. (2.33) in Eq. (2.21) yields the following cumulative distribution function:

$$F(t) = \int_0^t \frac{2}{\alpha^2}x\mathrm{e}^{-\left(\frac{x}{\alpha}\right)^2}\,\mathrm{d}x = 1 - \mathrm{e}^{-\left(\frac{t}{\alpha}\right)^2}\,. \tag{2.34}$$

2.5.4 Weibull Distribution

This continuous random variable distribution is named after W. Weibull, a Swedish mechanical engineering professor, and it can be used to represent many different physical phenomena [13]. The distribution probability density function is expressed by

$$f(t) = \frac{\theta}{\alpha^\theta} t^{\theta-1} \mathrm{e}^{-\left(\frac{t}{\alpha}\right)^\theta}, \quad \text{for } t \geq 0,\ \alpha > 0,\ \theta > 0\,, \tag{2.35}$$

where

θ and α are the distribution shape and scale parameters, respectively.

By substituting Eq. (2.35) into Eq. (2.21), we get the following equation for the cumulative distribution function:

$$F(t) = \int_0^t \frac{\theta}{\alpha^\theta} x^{\theta-1} \mathrm{e}^{-\left(\frac{x}{\alpha}\right)^\theta} \mathrm{d}x = 1 - \mathrm{e}^{-\left(\frac{t}{\alpha}\right)^\theta} \tag{2.36}$$

It is to be noted that both exponential and Rayleigh distributions are the special cases of Weibull distribution for $\theta = 1$ and $\theta = 2$, respectively.

2.5.5 Normal Distribution

This is one of the most widely used continuous random variable distributions and is also known as the Gaussian distribution after Carl Friedrich Gauss (1777 – 1855). However, the distribution was actually discovered by De Moivre in 1733 [11].

The probability density function of the distribution is defined by

$$f(t) = \frac{1}{\sigma\sqrt{2\pi}} \exp\left[-\frac{(t-\mu)^2}{2\sigma^2}\right], \quad \text{for } -\infty < t < +\infty\,, \tag{2.37}$$

where

μ is the distribution mean,
σ is the distribution standard deviation.

Using Eq. (2.37) in Eq. (2.21) yields the following equation for the cumulative distribution function:

$$F(t) = \frac{1}{\sigma\sqrt{2\pi}} \int_{-\infty}^{t} \exp\left[-\frac{(x-\mu)^2}{2\sigma^2}\right] \mathrm{d}x\,. \tag{2.38}$$

2.5.6 Lognormal Distribution

This is another continuous random variable distribution and is often used to represent failed equipment repair times. The distribution probability density function is expressed by

$$f(t) = \frac{1}{t\theta\sqrt{2\pi}} \exp\left[-\frac{(lnt - m)^2}{2\theta^2}\right], \quad \text{for } t \geq 0, \tag{2.39}$$

where

m and θ are the distribution parameters.

Using Eq. (2.39) in Eq. (2.21) yields the following cumulative distribution function:

$$F(t) = \frac{1}{\theta\sqrt{2\pi}} \int_{-\infty}^{t} \frac{1}{x} \exp\left[-\frac{(lnx - m)^2}{2\theta^2}\right] \mathrm{d}x. \tag{2.40}$$

2.6 Solving Differential Equations Using Laplace Transforms

Sometimes mining equipment reliability, maintainability, and safety studies may require finding solutions to a system of linear first-order differential equations. Under such circumstances, the application of Laplace transforms has proven to be a very effective approach. The following example demonstrates the application of Laplace transforms to finding solutions to a set of linear first-order differential equations describing a mining system:

Example 2.6

Assume that an engineering system used in mines can be, at any time t, in either of the three distinct states: working normally, failed in open mode, or failed in short mode. The following three linear first-order differential equations describe the mining system:

$$\frac{\mathrm{d}P_{\mathrm{w}}(t)}{\mathrm{d}t} + (\lambda_{\mathrm{om}} + \lambda_{\mathrm{sm}})P_{\mathrm{w}}(t) = 0, \tag{2.41}$$

$$\frac{\mathrm{d}P_{\mathrm{om}}(t)}{\mathrm{d}t} - \lambda_{\mathrm{om}}P_{\omega}(t) = 0, \tag{2.42}$$

$$\frac{\mathrm{d}P_{\mathrm{sm}}(t)}{\mathrm{d}t} - \lambda_{\mathrm{sm}}P_{\omega}(t) = 0, \tag{2.43}$$

where

$P_j(t)$ is the probability that the mining system is in state j at time t, $j = \mathrm{w}$ (working normally), $j = \mathrm{om}$ (failed in open mode), and $j = \mathrm{sm}$ (failed in short mode),

λ_{sm} is the mining system constant short mode failure rate,

λ_{om} is the mining system constant open mode failure rate.

At time $t = 0$, $P_{\mathrm{w}}(0) = 1$, $P_{\mathrm{om}}(0) = 0$, and $P_{\mathrm{sm}}(0) = 0$.

Find solutions to differential Eqs. (2.41)–(2.43) using Laplace transforms.

Using Table 2.1, Eqs. (2.41)–(2.43) and the given initial conditions we get

$$sP_{\mathrm{w}}(s) - 1 + (\lambda_{\mathrm{om}} + \lambda_{\mathrm{sm}})P_{\mathrm{w}}(s) = 0\,, \tag{2.44}$$

$$sP_{\mathrm{om}}(s) - \lambda_{\mathrm{om}}P_{\mathrm{w}}(s) = 0\,, \tag{2.45}$$

$$sP_{\mathrm{sm}}(s) - \lambda_{\mathrm{sm}}P_{\mathrm{w}}(s) = 0\,. \tag{2.46}$$

Solving Eqs. (2.44)–(2.46) we obtain

$$P_{\mathrm{w}}(s) = \frac{1}{s + \lambda_{\mathrm{om}} + \lambda_{\mathrm{sm}}}\,, \tag{2.47}$$

$$P_{\mathrm{om}}(s) = \frac{\lambda_{\mathrm{om}}}{s(s + \lambda_{\mathrm{om}} + \lambda_{\mathrm{sm}})}\,, \tag{2.48}$$

$$P_{\mathrm{sm}}(s) = \frac{\lambda_{\mathrm{sm}}}{s(s + \lambda_{\mathrm{om}} + \lambda_{\mathrm{sm}})}\,. \tag{2.49}$$

Taking the inverse Laplace transforms of Eqs. (2.47)–(2.49) we get

$$P_{\mathrm{w}}(t) = \mathrm{e}^{-(\lambda_{\mathrm{om}} + \lambda_{\mathrm{sm}})t}\,, \tag{2.50}$$

$$P_{\mathrm{om}}(t) = \frac{\lambda_{\mathrm{om}}}{\lambda_{\mathrm{om}} + \lambda_{\mathrm{sm}}}\left[1 - \mathrm{e}^{-(\lambda_{\mathrm{om}} + \lambda_{\mathrm{sm}})t}\right]\,, \tag{2.51}$$

and

$$P_{\mathrm{sm}}(t) = \frac{\lambda_{\mathrm{sm}}}{\lambda_{\mathrm{om}} + \lambda_{\mathrm{sm}}}\left[1 - \mathrm{e}^{-(\lambda_{\mathrm{om}} + \lambda_{\mathrm{sm}})t}\right]\,. \tag{2.52}$$

Thus, Eqs. (2.50)–(2.52) are the solutions to differential Eqs. (2.41)–(2.43).

2.7 Problems

1. A mining equipment manufacturing company inspected eight identical mining systems and found 10, 11, 2, 20, 6, 9, 4, and 5 defects in each system. Calculate the average number of defects per mining system (*i.e.*, arithmetic mean of the data set).
2. Calculate the mean deviation of the data values given in the above problem (*i.e.*, Problem 1).
3. Discuss the history of probability.
4. Define standard deviation.
5. Prove Eq. (2.12).
6. Discuss five important properties of probability.
7. Mathematically, define probability.
8. Determine expected value of Eq. (2.31).

9. Write down probability density functions for the following statistical distributions:
   - Normal distribution
   - Weibull distribution
10. Obtain the Laplace transform for the following function:

$$f(t) = t\,\mathrm{e}^{-\lambda t}\,, \tag{2.53}$$

where

t is the time variable,
λ is a constant.

References

1. Eves, H.: An Introduction to the History of Mathematics. Holt, Rinehart, and Winston, New York (1976)
2. Owen, D.B. (ed.): On the History of Statistics and Probability. Dekker, New York (1976)
3. Shooman, M.L.: Probabilistic Reliability: An Engineering Approach. McGraw-Hill, New York (1968)
4. Spiegel, M.R.: Probability and Statistics. McGraw-Hill, New York (1975)
5. Spiegel, M.R.: Statistics. McGraw-Hill, New York (1961)
6. Lipschutz, S.: Set Theory. McGraw-Hill, New York (1964)
7. Fault Tree Handbook, Report No. NUREG-0492, US Nuclear Regulatory Commission, Washington, DC (1981)
8. Mann, N.R., Schafer, R.E., Singpurwalla, N.D.: Methods for Statistical Analysis of Reliability and Life Data. Wiley, New York (1974)
9. Spiegel, M.R.: Laplace Transforms. McGraw-Hill, New York (1965)
10. Oberhettinger, F., Badic, L.: Tables of Laplace Transforms. Springer, New York (1973)
11. Patel, J.K., Kapadia, C.H., Owen, D.B.: Handbook of Statistical Distributions. Dekker, New York (1976)
12. Davis, D.J.: An analysis of some failure data. J. Am. Stat. Assoc. 113–150 (June 1952)
13. Weibull, W.: A statistical distribution function of wide applicability. J. Appl. Mech. 18, 293–297 (1951)

Chapter 3
Introduction to Reliability, Maintainability, and Safety

3.1 Introduction

The history of the reliability field may be traced back to the early 1930s when probability concepts were applied to problems associated with electric power generation [1–3]. However, generally the real beginning of the reliability field is regarded as World War II, when Germans applied basic reliability concepts to improve the reliability of their V1 and V2 rockets. Today, reliability engineering is a well-developed discipline and has branched out into specialized areas such as software reliability, mechanical reliability, and human reliability. A detailed history of the reliability field is available in Ref. [4].

The beginning of the maintainability field may be traced back to 1901 to the United States Army Signal Corps contract for the development of the Wright brothers' airplane. In this document, it was clearly stated that the aircraft should be "simple to operate and maintain" [5]. The first commercially available book entitled *Electronic Maintainability* appeared in 1960, and in the latter part of the 1960s many military documents on maintainability were published by the United States Department of Defense [6–8]. A detailed history of the maintainability field is given in Refs. [9, 10].

The history of the safety field goes back to 1868, when a patent was awarded for a barrier safeguard in the United States [11]. In 1893, the Railway Safety Act was passed by the U.S. Congress, and in 1931 the first commercially available book, *Industrial Accident Prevention*, was published [12]. A detailed history of safety is available in Ref. [13].

This chapter presents various important introductory aspects of reliability, maintainability, and safety considered useful for the mining industry.

3.2 Need for Reliability and Bathtub Hazard Rate Curve

Today reliability has become an important factor during the design phase of engineering systems because our daily lives and schedules are increasingly becoming more dependent than ever before on the satisfactory functioning of such systems. Some examples of these systems are aircraft, trains, automobiles, space satellites, and computers. Some of the specific factors responsible for the consideration of reliability in system/product design are the insertion of reliability-related clauses in design specifications, product/system complexity and sophistication, high acquisition cost, competition, public demand, past system/product failures, loss of prestige, and the increasing number of reliability-/safety-/quality-related lawsuits.

The bathtub hazard rate curve, shown in Fig. 3.1, is widely used to represent the failure rate of various types of engineering items. As shown in Fig. 3.1, the curve is divided into three regions: burn-in period, useful-life period, and wear-out period. During the burn-in period the item hazard rate or time-dependent failure rate decreases with time t. Some of the reasons for the occurrence of failures during this period are poor quality control, substandard materials and workmanship, poor manufacturing methods, inadequate debugging, poor processes, and human error [14].

During the useful-life period the item hazard rate remains constant. Some of the reasons for the occurrence of failures during this period are low safety factors, undetectable defects, natural failures, abuse, higher-than-expected random stress, and human error. Finally, during the wear-out period the item hazard rate increases with time t. Some of the causes for the occurrence of failures during this period are

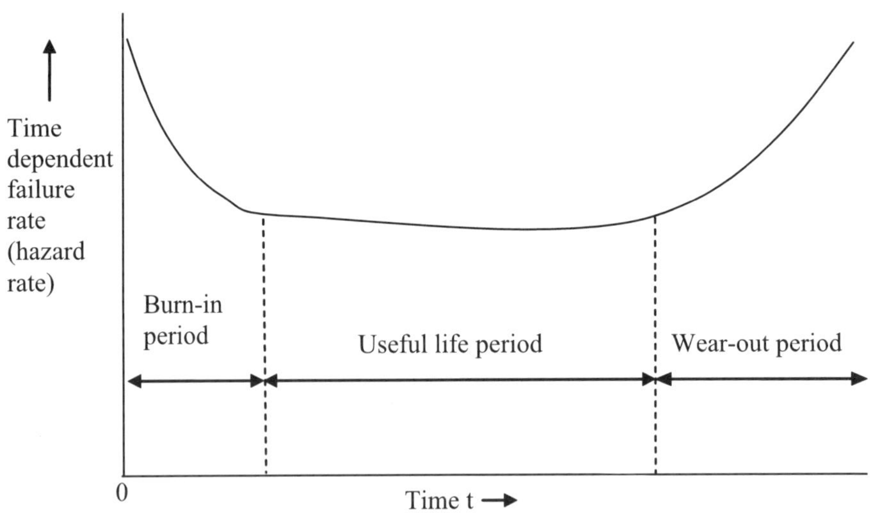

Fig. 3.1 Bathtub hazard rate curve

wear caused by friction, poor maintenance, wear caused by aging, corrosion and creep, incorrect overhaul practices, and short designed-in life of the item.

3.3 General Reliability, Hazard Rate, and Mean Time to Failure Functions

Many general functions are frequently used in performing reliability analysis. Three of these functions are presented below.

3.3.1 General Reliability Function

This is expressed by

$$R(t) = \mathrm{e}^{-\int_0^t \lambda(t)\,\mathrm{d}t}\,, \tag{3.1}$$

where

$R(t)$ is the reliability at time t,
$\lambda(t)$ is the hazard rate or time-dependent failure rate.

Equation (3.1) is the general expression for the reliability function. It is used to obtain the reliability of an item whose times to failure are described by statistical distributions such as exponential, Rayleigh, Weibull, and normal.

Example 3.1

Assume that the times to failure of a piece of mining equipment are exponentially distributed. Thus, the equipment's hazard rate is given by

$$\lambda(t) = \lambda\,, \tag{3.2}$$

where

λ is the mining equipment constant failure rate.

Obtain an expression for the equipment reliability function.

Substituting Eq. (3.2) into Eq. (3.1) we get

$$R(t) = \mathrm{e}^{-\int_0^t \lambda\,\mathrm{d}t} = \mathrm{e}^{-\lambda t}\,. \tag{3.3}$$

Equation (3.3) is the reliability function for the piece of mining equipment.

3.3.2 Hazard Rate Function

The hazard rate function is expressed by

$$\lambda(t) = \frac{f(t)}{R(t)} \tag{3.4}$$

or

$$\lambda(t) = -\frac{1}{R(t)} \cdot \frac{\mathrm{d}R(t)}{\mathrm{d}t}, \tag{3.5}$$

where

$f(t)$ is the failure (or probability) density function.

Example 3.2

Prove Eq. (3.2) using Eqs. (3.3) and (3.5).

Thus, substituting Eq. (3.3) into Eq. (3.5) we get

$$\lambda(t) = -\frac{1}{\mathrm{e}^{-\lambda t}} \cdot \frac{\mathrm{d}\,\mathrm{e}^{-\lambda t}}{\mathrm{d}t} = \lambda\,. \tag{3.6}$$

Equations (3.2) and (3.6) are identical.

3.3.3 Mean Time to Failure

Mean time to failure can be obtained using any of the following formulas [15]:

$$MTTF = \int_0^\infty R(t)\,\mathrm{d}t \tag{3.7}$$

or

$$MTTF = \lim_{s \to 0} R(s) \tag{3.8}$$

or

$$MTTF = \int_0^\infty t f(t)\,\mathrm{d}t\,, \tag{3.9}$$

where

$MTTF$ is the mean time to failure,
s is the Laplace transform variable,
$R(s)$ is the Laplace transform of the reliability function, $R(t)$.

Example 3.3

Using Eq. (3.3), obtain an expression for the mining equipment mean time to failure.
Substituting Eq. (3.3) into Eq. (3.7) we get

$$MTTF = \int_0^{\infty} \mathrm{e}^{-\lambda t}\,\mathrm{d}t = \frac{1}{\lambda}\,. \tag{3.10}$$

Equation (3.10) is the expression for the mining equipment mean time to failure.

3.4 Reliability Networks

A system can form various types of networks or configurations in performing reliability analysis. Some commonly occurring configurations are presented below.

3.4.1 Series Configuration

This is probably the most widely occurring configuration in engineering systems; it is depicted by the block diagram shown in Fig. 3.2. Each block in the diagram denotes a component or unit. In this arrangement or configuration, all the units must operate normally for the successful operation of the system (*i.e.*, the series system).

If we let X_j denote the event that the jth unit in Fig. 3.2 is successful, then the reliability of the series configuration/system is given by

$$R_s = P(X_1X_2X_3 \ldots X_m)\,, \tag{3.11}$$

where

R_s is the series system or configuration reliability,
$P(X_1X_2X_3 \ldots X_m)$ is the occurrence probability of success events $X_1X_2X_3 \ldots$ and X_m.

For independent units, Eq. (3.11) becomes

$$R_s = P(X_1)P(X_2)P(X_3) \ldots P(X_m)\,, \tag{3.12}$$

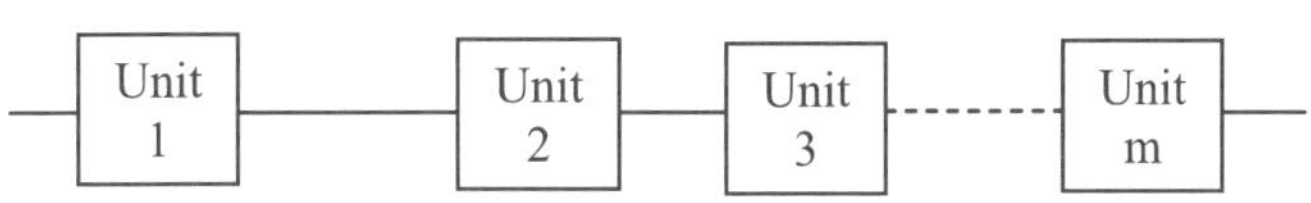

Fig. 3.2 A series configuration with m units

where

$P(X_j)$ is the probability of occurrence of success event X_j, for $j = 1,2,3,\ldots,m$.

If we let $R_j = P(X_j)$ for $j = 1,2,3,\ldots,m$ in Eq. (3.12), the equation becomes

$$R_s = \prod_{j=1}^{m} R_j \,, \tag{3.13}$$

where

R_j is the unit j reliability; for $j = 1,2,3,\ldots,m$.

For constant failure rate, λ_j, of unit j, using Eq. (3.1) the reliability of unit j is given by

$$R_j(t) = \mathrm{e}^{-\int_0^t \lambda_j \,\mathrm{d}t} = \mathrm{e}^{-\lambda_j t} \,, \tag{3.14}$$

where

$R_j(t)$ is the reliability of unit j at time t.

Substituting Eq. (3.14) into Eq. (3.13) we get

$$R_s(t) = \mathrm{e}^{-\sum_{j=1}^{m} \lambda_j t} \,, \tag{3.15}$$

where

$R_s(t)$ is the reliability of the series system at time t.

Inserting Eq. (3.15) into Eq. (3.7) we get

$$MTTF_s = \int_0^\infty \mathrm{e}^{-\sum_{j=1}^{m} \lambda_j t} \,\mathrm{d}t = \frac{1}{\sum_{j=1}^{m} \lambda_j} \,, \tag{3.16}$$

where

$MTTF_s$ is the series system mean time to failure.

Example 3.4

Assume that a mining system is composed of five independent and identical subsystems in series. The constant failure rate of each subsystem is 0.0006 failures per hour. Calculate the mining system mean time to failure and reliability for a 100-h mission.

Using the data values given in Eq. (3.16) yields

$$MTTF_s = \frac{1}{5(0.0006)} = 333.3\,\mathrm{h} \,.$$

Substituting the specified data into Eq. (3.15) we get

$$R_s(100) = e^{-5(0.0006)(100)} = 0.7408\,.$$

Thus, the mining system mean time to failure and reliability are 333.3 h and 0.7408, respectively.

3.4.2 Parallel Configuration

In this case, all m units are active and at least one of these units must operate normally for the successful operation of the system. The block diagram of an "m" unit parallel configuration/system is shown in Fig. 3.3; each block in the diagram represents a unit.

If we let $\bar{X}_j$ represent the event that the jth unit is unsuccessful, then the failure probability of the parallel system/configuration is given by

$$F_p = P(\bar{X}_1\bar{X}_2\ldots\bar{X}_m)\,, \tag{3.17}$$

where

F_p is the parallel system/configuration reliability,
$P(\bar{X}_1\bar{X}_2\ldots\bar{X}_m)$ is the occurrence probability of failure events $\bar{X}_1, \bar{X}_2, \ldots, \bar{X}_m$.

For independent units, Eq. (3.17) becomes

$$F_p = P(\bar{X}_1)P(\bar{X}_2)\ldots P(\bar{X}_m)\,, \tag{3.18}$$

where

$P(\bar{X}_j)$ is the occurrence probability of failure event $\bar{X}_j$; for $j = 1, 2, 3, \ldots, m$.

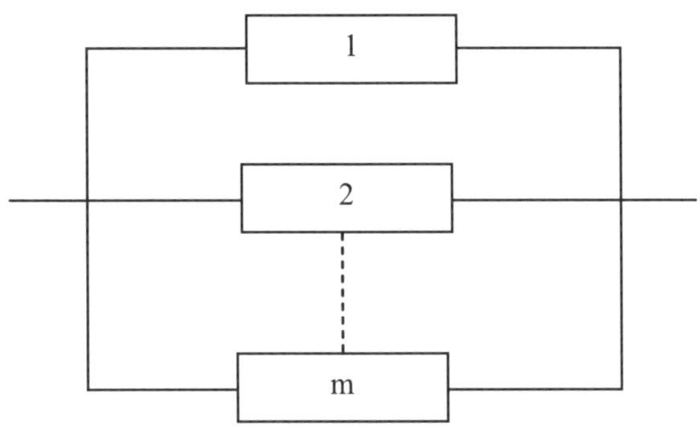

Fig. 3.3 Block diagram of a parallel configuration containing m units

If we let $F_j = P(\bar{X}_j)$ for $j = 1,2,\ldots,m$ in Eq. (3.18), the equation becomes

$$F_p = \prod_{j=1}^{m} F_j \,, \tag{3.19}$$

where

F_j is the failure probability of unit j; for $j = 1,2,\ldots,m$.

Subtracting Eq. (3.19) from unity we obtain

$$R_p = 1 - \prod_{j=1}^{m} F_j \,, \tag{3.20}$$

where

R_p is the parallel system/configuration reliability.

For the constant failure rate, λ_j, of unit j, subtracting Eq. (3.14) from unity; then, substituting it into Eq. (3.20) yields

$$R_p(t) = 1 - \prod_{j=1}^{m} \left(1 - e^{-\lambda_j t}\right) \,, \tag{3.21}$$

where

$R_p(t)$ is the parallel system/configuration reliability at time t.

For identical units, using Eq. (3.7) and (3.21) we get

$$MTTF_p = \int_0^{\infty} \left[1 - \left(1 - e^{-\lambda t}\right)^m\right] dt = \frac{1}{\lambda} \sum_{j=1}^{m} \frac{1}{j} \,, \tag{3.22}$$

where

λ is the unit constant failure rate,
$MTTF_p$ is the parallel system/configuration mean time to failure.

Example 3.5

A mining system is composed of three independent and identical units in parallel, and their constant failure rates are 0.0005 failures per hour. Calculate the system mean time to failure and reliability for a 200-h mission.

Substituting the specified data values into Eq. (3.22) we get

$$MTTF_p = \frac{1}{(0.0005)} \left(1 + \frac{1}{2} + \frac{1}{3}\right) = 3666.7\,\text{h} \,.$$

Using the given data values in Eq. (3.21) yields

$$R_{\mathrm{p}}(200) = 1 - \left(1 - \mathrm{e}^{-(0.0005)(200)}\right)^2 = 0.9909 .$$

Thus, the mining system mean time to failure and reliability are 3666.7 h and 0.9909, respectively.

3.4.3 *k*-out-of-*m* Configuration

In this case m number of units are active and at least k units must operate normally for the system success. The parallel and series configurations are special cases of this configuration for $k = 1$ and $k = m$, respectively.

Using the binomial distribution, for independent and identical units, we write down the following expression for the k-out-of-m configuration reliability:

$$R_{k/m} = \sum_{j=k}^{m} \binom{m}{j} R^j (1-R)^{m-j} , \tag{3.23}$$

where

$$\binom{m}{j} = \frac{m!}{(m-j)!j!} , \tag{3.24}$$

$R_{k/m}$ is the k-out-of-m configuration/system reliability,
R is the unit reliability.

For constant failure rates of units, using Eqs. (3.3) and (3.23) we get

$$R_{k/m}(t) = \sum_{j=k}^{m} \binom{m}{j} \mathrm{e}^{-j\lambda t} \left(1 - \mathrm{e}^{-\lambda t}\right)^{m-j} , \tag{3.25}$$

where

$R_{k/m}(t)$ is the k-out-of-m configuration/system reliability at time t,
λ is the unit constant failure rate.

Substituting Eq. (3.25) into Eq. (3.7) we get

$$MTTF_{k/m} = \int_0^{\infty} \left[\sum_{j=k}^{m} \binom{m}{j} \mathrm{e}^{-j\lambda t} \left(1 - \mathrm{e}^{-\lambda t}\right)^{m-j} \right] \mathrm{d}t = \frac{1}{\lambda} \sum_{j=k}^{m} \frac{1}{j} , \tag{3.26}$$

where

$MTTF_{k/m}$ is the mean time to failure of the k-out-of-m configuration/system.

Example 3.6

A mining system is composed of three independent and identical units operating in parallel. At least two of these units must operate normally for the system to succeed. Calculate the mining system mean time to failure if the unit failure rate is 0.001 failures per hour.

Substituting the specified data values into Eq. (3.26) we get

$$MTTF_{2/3} = \frac{1}{(0.001)} \sum_{j=2}^{3} \frac{1}{j} = 833.3\,\text{h}\,.$$

Thus, the mining equipment mean time to failure is 833.3 h.

3.4.4 Standby System

In this case, the system is composed of $(n+1)$ units and only one unit operates and the remaining n units are kept in their standby mode. As soon as the operating unit fails, the switching mechanism detects the failure and then turns on one of the n standby units. The system fails when all the n standby units fail.

For independent and identical units (*i.e.*, the operating plus standby units), perfect switching mechanism and standby units, the standby system reliability is expressed by

$$R_{sb}(t) = \sum_{j=0}^{n} \left[\left[\int_0^t \lambda(t)\,\mathrm{d}t \right]^j \mathrm{e}^{-\int_0^t \lambda(t)\,\mathrm{d}t} \right] / j!\,, \tag{3.27}$$

where

$R_{sb}(t)$	is the standby system reliability at time t,
$\lambda(t)$	is the unit time dependent failure rate,
n	is the number of standby units.

For constant unit failure rate (*i.e.*, $\lambda(t) = \lambda$), using Eq. (3.27), we get

$$R_{sb}(t) = \sum_{j=0}^{n} \left[(\lambda t)^j \mathrm{e}^{-\lambda t} \right] / j!\,. \tag{3.28}$$

Substituting Eq. (3.28) into Eq. (3.7) we obtain

$$MTTF_{sb} = \int_0^{\infty} \left[\sum_{j=0}^{n} (\lambda t)^j \mathrm{e}^{-\lambda t} / j! \right] \mathrm{d}t = \frac{n+1}{\lambda}\,, \tag{3.29}$$

where

$MTTF_{sb}$ is the standby system mean time to failure.

Example 3.7

A standby mining system is composed of three independent and identical units: one operating, two on standby. The unit constant failure rate is 0.0005 failures per hour. The standby unit turn-on mechanism is perfect and both the standby units remain as good as new in their standby mode. Calculate the mining system mean time to failure.

Substituting the given data values into Eq. (3.29) we get

$$MTTF_{\mathrm{sb}} = \frac{(2+1)}{(0.0005)} = 6000\,\mathrm{h}\,.$$

Thus, the standby mining system mean time to failure is 6000 h.

3.4.5 Bridge Configuration

This type of configuration also occurs in engineering systems. The block diagram of a bridge configuration is shown in Fig. 3.4. Each block in the diagram denotes a unit.

For independent units, the reliability of the Fig. 3.4 bridge configuration is expressed by [16]

$$\begin{aligned} R_{\mathrm{b}} = {} & 2R_1R_2R_3R_4R_5 + R_2R_3R_4 + R_1R_3R_5 + R_1R_4 + R_2R_5 - R_2R_3R_4R_5 \\ & - R_1R_2R_3R_4 - R_5R_1R_2R_3 - R_1R_3R_4R_5 - R_1R_2R_4R_5\,, \end{aligned} \tag{3.30}$$

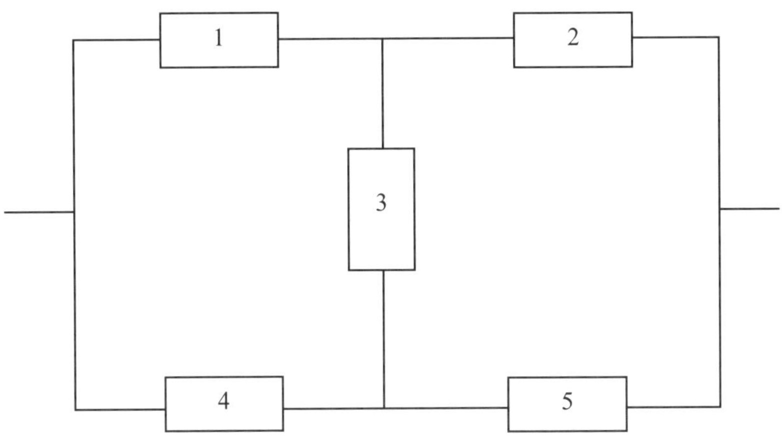

Fig. 3.4 A five-unit bridge configuration

where

R_b is the bridge configuration/system reliability.
R_j is the reliability of unit j; for $j = 1,2,3,\ldots,5$.

For identical units and constant failure rates of units, using Eq. (3.1) and (3.30) we get

$$R_b(t) = 2e^{-5\lambda t} - 5e^{-4\lambda t} + 2e^{-3\lambda t} + 2e^{-2\lambda t}, \tag{3.31}$$

where

$R_b(t)$ is the bridge configuration/system reliability at time t,
λ is the unit constant failure rate.

Substituting Eq. (3.31) into Eq. (3.7) we get

$$MTTF_b = \int_0^\infty \left[2e^{-5\lambda t} - 5e^{-4\lambda t} + 2e^{-3\lambda t} + 2e^{-2\lambda t}\right] dt = \frac{49}{60\lambda}, \tag{3.32}$$

where

$MTTF_b$ is the bridge configuration/system mean time to failure.

Example 3.8

Assume that five independent and identical units of a mining system form a bridge configuration. Calculate the bridge configuration reliability for a 100-h mission and mean time to failure, if the constant failure rate of each unit is 0.0004 failures per hour.

Using the specified data values in Eq. (3.31) yields

$$\begin{aligned} R_b(100) &= 2e^{-5(0.0004)(100)} - 5e^{-4(0.0004)(100)} \\ &\quad + 2e^{-3(0.0004)(100)} + 2e^{-2(0.0004)(100)} \\ &= 0.9968\,. \end{aligned}$$

Substituting the given data value into Eq. (3.32) we get

$$MTTF_b = \frac{49}{60(0.0004)} = 2041.67\,\text{h}\,.$$

Thus, the bridge configuration reliability and mean time to failure are 0.9968 and 2041.67 h, respectively.

3.5 Commonly Used Methods in Reliability Analysis

Over the years many methods have been developed to perform reliability analysis of engineering systems. These methods are particularly useful for analyzing engineering systems more complex than the ones forming the standard reliability configurations.

This section presents three of these methods considered useful for application in the mining industry [4, 17–19].

3.5.1 Failure Modes and Effect Analysis (FMEA)

This is one of the most widely used methods for performing reliability analysis of engineering systems and is basically a qualitative approach. It was developed in the early 1950s to assess the designs of flight control systems [20].

FMEA usually starts during the early phases of system design and is performed by following the seven steps shown in Fig. 3.5. Some of the questions asked during the performance of FMEA with respect to components/subsystems are as follows:

- What are the possible failure modes of the component/subsystem?
- What are the possible consequences of the failure mode?
- How is failure detected?
- How critical are the consequences?
- What are the effective safeguards against the failure in question?

Some of the important applications of FMEA are as follows [19]:

- To identify weak spots in design,
- To choose design alternatives during the early stages of design,
- To serve as a basis for design improvement action,
- To identify weak areas in design,
- To ensure the understanding of all possible failure modes and their anticipated effects,
- To choose design alternatives during the early stages of design, and
- To recommend appropriate test programs.

Additional information on FMEA is available in Ref. [4].

3.5.2 Markov Method

This is a widely used method to handle reliability and availability analyses of repairable systems. The approach proceeds by the enumeration of system states, and then the resulting system of differential equations are solved to obtain reliability-

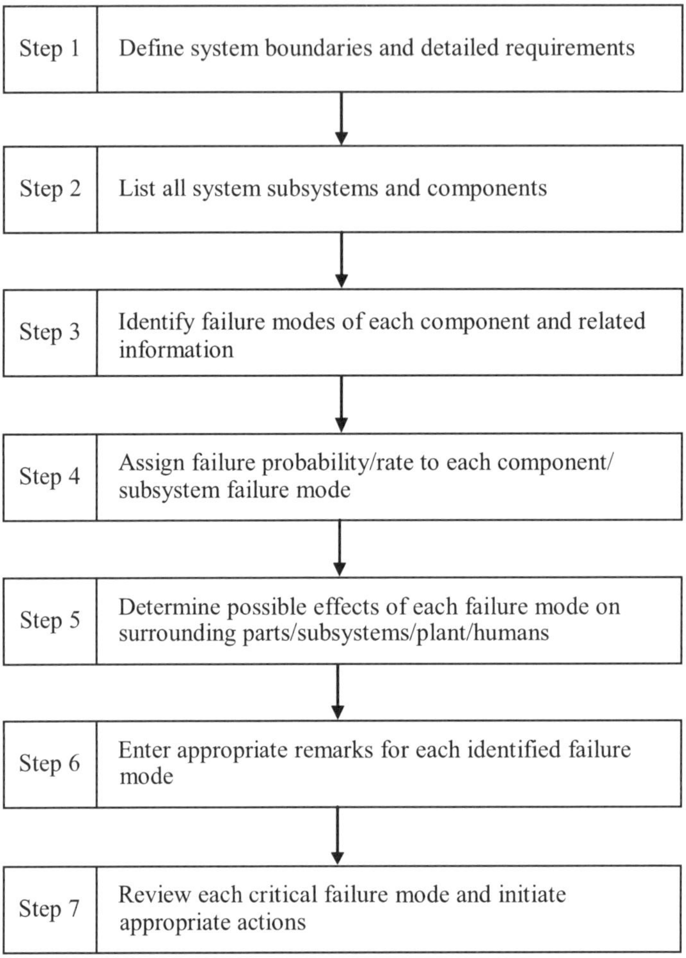

Fig. 3.5 Steps for performing FMEA

related measures. The following assumptions are associated with the Markov method [21]:

- All system transition rates (*i.e.*, failure and repair rates) are constant.
- All occurrences are independent of each other.
- The probability of transition from one system state to another in the finite time interval Δt is given by $\lambda \Delta t$, where λ is the transition rate (*e.g.*, system failure or repair rate) from one system state to another.
- The probability of more than one transition occurrence in finite time interval Δt from one system state to another is very small or negligible [*e.g.*, $(\lambda \Delta t)(\lambda \Delta t) \rightarrow 0$].

The application of the method is demonstrated by solving the following example.

Example 3.9

A mining system can either be in an operating state or a failed state, and its constant failure and repair rates are λ_m and μ_m, respectively. The system state-space diagram is shown in Fig. 3.6. The numerals in boxes denote system states. Obtain expressions for the mining system state probabilities using the Markov method.

Using Fig. 3.6 and the Markov method, we write down the following two equations [4, 21]:

$$P_0(t+\Delta t) = P_0(t)(1-\lambda_m \Delta t) + P_1(t)\mu_m \Delta t\ , \tag{3.33}$$

$$P_1(t+\Delta t) = P_1(t)(1-\mu_m \Delta t) + P_0(t)\lambda_m \Delta t\ , \tag{3.34}$$

where

λ_m	is the mining system constant failure rate,
μ_m	is the mining system constant repair rate,
t	is time,
$\lambda_m \Delta t$	is the probability of the mining system failure in finite time interval Δt.
$\mu_m \Delta t$	is the probability of the mining system repair in finite time interval Δt.
$(1-\lambda_m \Delta t)$	is the probability of no mining system failure in finite time interval Δt.
$(1-\mu_m \Delta t)$	is the probability of no mining system repair in finite time interval Δt.
$P_i(t+\Delta t)$	is the probability of the mining system being in state i at time $(t+\Delta t)$; for $i=0$ (operating normally), $i=1$ (failed).
$P_i(t)$	is the probability that the mining system is in state i at time t; for $i=0,1$.

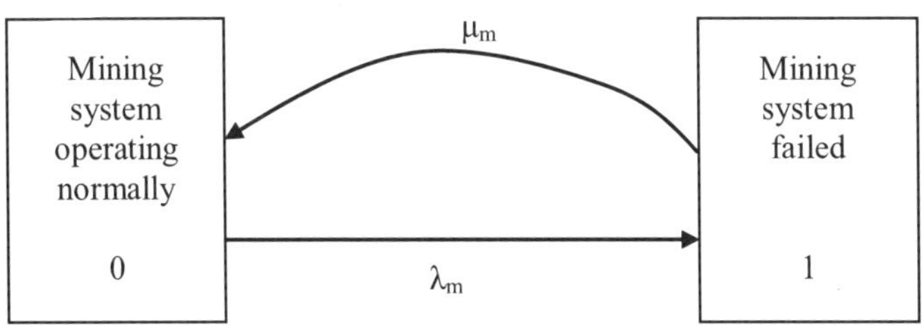

Fig. 3.6 Mining system state-space diagram

In the limiting case, Eqs. (3.32) and (3.33) become

$$\frac{dP_0(t)}{dt} + \lambda_m P_0(t) = P_1(t)\mu_m \,, \tag{3.35}$$

$$\frac{dP_1(t)}{dt} + \mu_m P_1(t) = P_0(t)\lambda_m \,. \tag{3.36}$$

At time $t = 0$, $P_0(0) = 1$ and $P_1(0) = 0$.

By solving Eqs. (3.35) and (3.36), we obtain

$$P_0(t) = \frac{\mu_m}{(\lambda_m + \mu_m)} + \frac{\lambda_m}{(\lambda_m + \mu_m)} e^{-(\lambda_m + \mu_m)t} \tag{3.37}$$

$$P_1(t) = \frac{\lambda_m}{(\lambda_m + \mu_m)} - \frac{\lambda_m}{(\lambda_m + \mu_m)} e^{-(\lambda_m + \mu_m)t} \tag{3.38}$$

Thus, Eqs. (3.37) and (3.38) are the expressions for the mining system state probabilities.

Example 3.10

Assume that in Example 3.9 the values of λ_m and μ_m are 0.002 failures per hour and 0.004 repairs per hour, respectively. Calculate the probabilities of the mining system operating normally and failed for a 100-h mission.

Substituting the given data values into Eqs. (3.37) and (3.38) we get

$$P_0(100) = \frac{0.004}{(0.002 + 0.004)} + \frac{0.002}{(0.002 + 0.004)} e^{-(0.002+0.004)(100)} = 0.8496$$

and

$$P_1(100) = \frac{0.002}{(0.002 + 0.004)} - \frac{0.002}{(0.002 + 0.004)} e^{-(0.002+0.004)(100)} = 0.1504 \,.$$

Thus, the probabilities of the mining system operating normally and failed are 0.8496 and 0.1504, respectively.

3.5.3 Fault Tree Analysis

This is another widely used method to perform reliability analysis of engineering systems. It was developed at the Bell Telephone Laboratories in the early 1960s to analyze the Minuteman Launch Control System with respect to reliability and safety [22].

Although this method makes use of a large number of symbols, the four commonly used symbols are shown in Fig. 3.7 [4, 22].

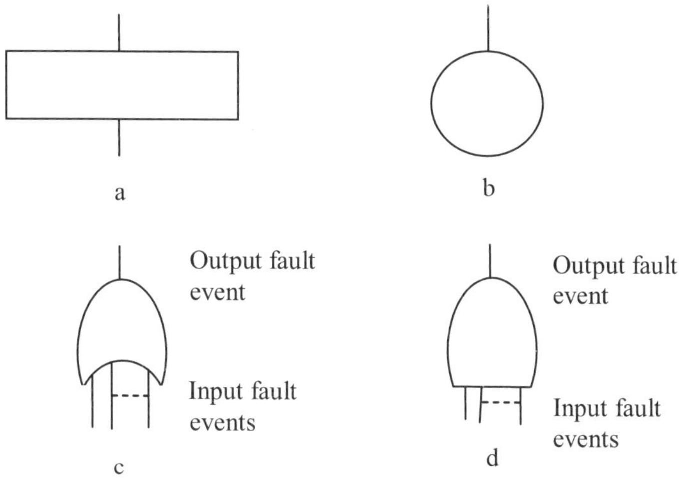

Fig. 3.7a–d Commonly used fault tree symbols. **a** Rectangle. **b** Circle. **c** OR gate. **d** AND gate

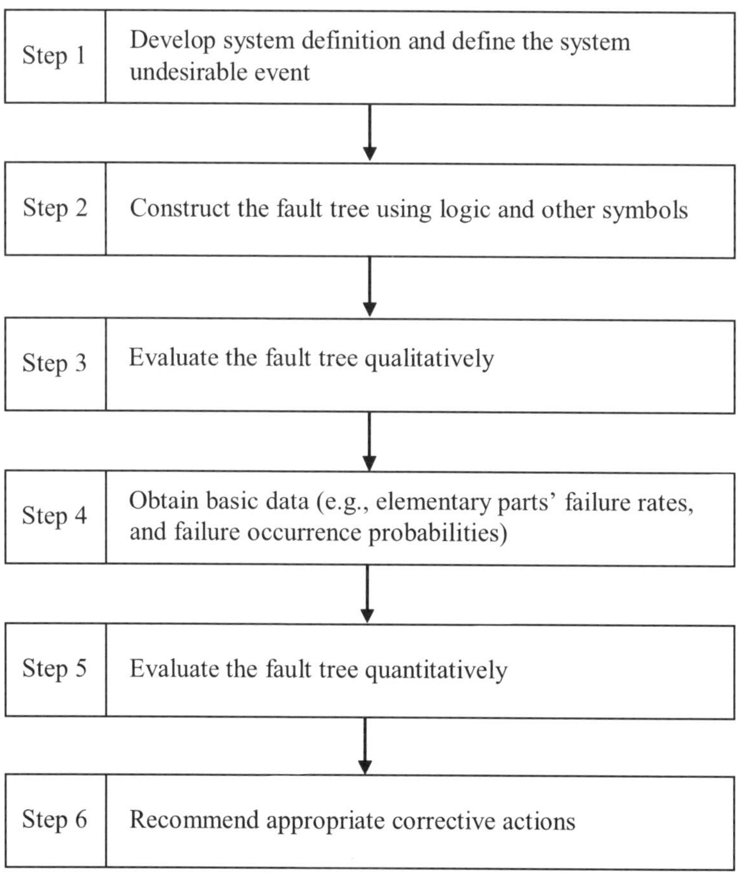

Fig. 3.8 Steps for developing a fault tree

The Fig. 3.7 symbols are defined below.

- **Rectangle.** This represents a resultant event that results from the combination of fault events through the input of a logic gate.
- **Circle.** This denotes a basic fault event or the failure of an elementary part.
- **OR gate.** This denotes that an output fault event occurs if one or more of the input fault events occur.
- **AND gate.** This denotes that an output fault event occurs only if all the input fault events occur.

Six basic steps used to develop a fault tree are shown in Fig. 3.8. The fault tree construction starts from the undesirable event known as the top event and then successively asking the question "How could this event occur?" until reaching the desirable basic fault events.

In order to estimate the probability of occurrence of the top event, it is essential to estimate the probability of occurrence of the logic gates' output fault events. Thus, equations to estimate the probability of occurrence of OR and AND logic gates' output fault events are presented below.

OR Gate

The probability of occurrence of an OR gate's output fault event is given by [4]

$$P(X_o) = 1 - \prod_{j=1}^{m} (1 - P(X_j)) \,, \tag{3.39}$$

where

m is the number of input fault events;
$P(X_o)$ is the probability of occurrence of OR gate's output fault event X_o;
$P(X_j)$ is the probability of occurrence of input fault event X_j;
for $j = 1, 2, 3, \ldots, m$.

AND Gate

The probability of occurrence of an AND gate's output fault event is given by

$$P(X_a) = \prod_{j=1}^{m} P(X_j) \,, \tag{3.40}$$

where

$P(X_a)$ is the probability of occurrence of AND gate's output fault event X_a.

The application of the fault tree analysis method is demonstrated through the following example.

Example 3.11

A windowless room has four lightbulbs controlled by a single switch that can only fail to close. Using Fig. 3.7 symbols, construct a fault tree for an undesirable event (top event): dark room. Furthermore, calculate the probability of occurrence of the undesirable event, if all basic fault events occur independently and their occurrence probability (*i.e.*, each event's) is 0.15.

The fault tree shown in Fig. 3.9 was developed using Fig. 3.7 symbols. Single capital letters in parentheses in circles and rectangles denote corresponding fault events. Probabilities of occurrence of events T, A, and B in Fig. 3.9 are calculated below.

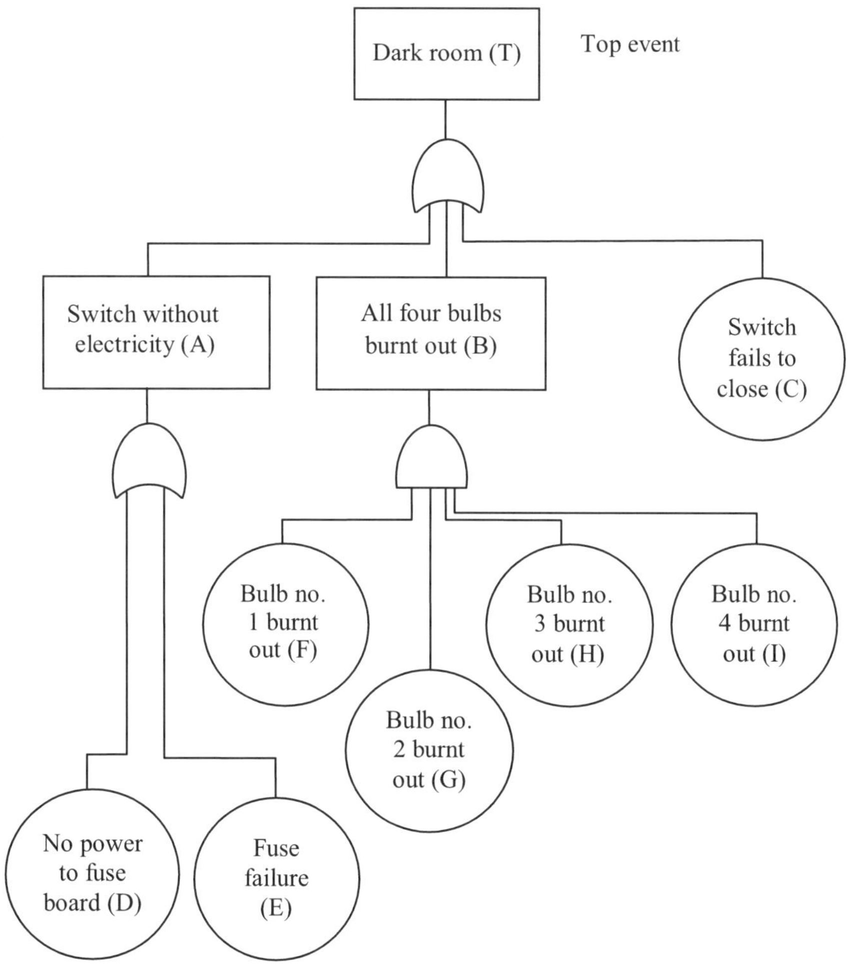

Fig. 3.9 Fault tree for the dark room undesirable event

Substituting the given data values into Eq. (3.39), we get the following probability of occurrence of fault event A:

$$
\begin{aligned}
P(A) &= 1-(1-P(D))(1-P(E)) \\
&= 1-(1-0.15)(1-0.15) \\
&= 0.2775\,,
\end{aligned}
$$

where

$P(A)$ is the probability of having the switch without electricity,
$P(D)$ is the probability of having no power to fuse board,
$P(E)$ is the probability of the fuse failure.

Using the specified data values in Eq. (3.40) we get

$$
\begin{aligned}
P(B) &= P(F)P(G)P(H)P(I) \\
&= (0.15)(0.15)(0.15)(0.15) \\
&= 0.0005\,,
\end{aligned}
$$

where

$P(B)$ is the probability of having all four bulbs burn out,
$P(F)$ is the probability of bulb number 1 burning out,
$P(G)$ is the probability of bulb number 2 burning out,
$P(H)$ is the probability of bulb number 3 burning out,
$P(I)$ is the probability of bulb number 4 burning out.

Substituting the calculated and given values into Eq. (3.39) we obtain

$$
\begin{aligned}
P(T) &= 1-(1-P(A))(1-P(B))(1-P(C)) \\
&= 1-(1-0.2775)(1-0.0005)(1-0.15) \\
&= 0.3862\,,
\end{aligned}
$$

where

$P(T)$ is the probability of occurrence of the top event: dark room,
$P(C)$ is the probability of having switch fails to close.

Thus, the probability of occurrence of the dark room undesirable event is 0.3862.

3.6 Need for Maintainability and Maintainability Versus Reliability

The need for maintainability is becoming increasingly important because of the alarmingly high operating and support costs of engineering systems. For example, each year over $300 billion is being spent by American manufacturers on plant maintenance and operations [23]. Thus, some of the main objectives of applying maintainability principles to engineering systems are to reduce projected maintenance costs and time, to use maintainability data to estimate system/equipment availability/unavailability, and to determine labor-hours and other related resources needed to perform the projected maintenance.

Maintainability is a built-in design and installation characteristic that provides the resulting system/equipment with an inherent ability to be maintained, leading to lower maintenance costs, required skill levels, required tools and equipment, required man-hours, and better mission availability. In contrast, reliability is a design characteristic that leads to the durability of the system as it performs its specified mission according to a stated condition and time period. It is accomplished through various measures including selecting optimum engineering principles, controlling processes, satisfactory component sizing, and testing.

Some of the specific principles of maintainability (and of the corresponding reliability) are reducing life cycle maintenance costs (maximize the use of standard parts), reducing or eliminating the need for maintenance (minimizing stress on components and parts), reducing the amount, frequency, and complexity of required maintenance tasks (use fewer components for performing multiple functions), providing for maximum interchangeability (provide fail-safe designs), and reducing mean time to repair (design for simplicity) [24].

3.7 Maintainability Functions

There are many maintainability functions depending on the time to repair distribution. The maintainability function is used to predict the probability that a repair, starting at time $t = 0$, will be accomplished within time t. Mathematically, the maintainability function is defined by [5]

$$m(t) = \int_0^t f_r(t)\,\mathrm{d}t\,, \tag{3.41}$$

where

$m(t)$ is the maintainability function,
t is time,
$f_r(t)$ is the repair time probability density function.

Maintainability functions for two commonly occurring repair time distributions are presented below [26, 27]. Similarly, one can obtain maintainability functions for other repair time distributions.

3.7.1 Maintainability Function I: Exponential Distribution

In this case times to repair are exponentially distributed and are defined by

$$f_r(t) = \mu\, \mathrm{e}^{-\mu t}\,, \tag{3.42}$$

where

t is the variable repair time,
μ is the constant repair rate or reciprocal of the mean time to repair (MTTR).

Using Eq. (3.42) in Eq. (3.41) yields the following maintainability function for the exponential distribution:

$$m_{\mathrm{e}}(t) = \int_0^t \mu\, \mathrm{e}^{-\mu t}\, \mathrm{d}t = 1 - \mathrm{e}^{-\left(\frac{1}{MTTR}\right)t}\,, \tag{3.43}$$

where

$$\mu = \frac{1}{MTTR}\,,$$

$m_{\mathrm{e}}(t)$ is the maintainability function for exponential distribution.

Example 3.12

Assume that the repair times of pieces of mining equipment are exponentially distributed with a mean value (*i.e.*, MTTR) of 10 h. Calculate the probability of accomplishing a repair within 15 h.

Inserting the given data values into Eq. (3.43) we get

$$m_{\mathrm{e}}(15) = 1 - \mathrm{e}^{-\left(\frac{1}{10}\right)(15)} = 0.7769\,.$$

Thus, there is a 77.69% chance that the mining equipment repair will be accomplished within 15 h.

3.7.2 Maintainability Function II: Weibull Distribution

In this case, times to repair are Weibull distributed and are defined by

$$f_r(t) = \frac{b}{\alpha^b} t^{b-1}\, \mathrm{e}^{-\left(\frac{t}{\alpha}\right)^b}\,, \tag{3.44}$$

where

b is the distribution shape parameter,
α is the distribution scale parameter.

Inserting Eq. (3.44) into Eq. (3.41) we get

$$m_{\mathrm{w}}(t) = \int_0^t \frac{b}{\alpha^b} t^{b-1} \mathrm{e}^{-\left(\frac{t}{\alpha}\right)^b} \mathrm{d}t = 1 - \mathrm{e}^{-\left(\frac{t}{\alpha}\right)^b}, \tag{3.45}$$

where

$m_{\mathrm{w}}(t)$ is the maintainability function for Weibull distribution.

3.8 Maintainability Design Factors and Maintainability Analysis Tools

There are numerous goals of maintainability design including increasing ease of maintenance, minimizing preventive and corrective maintenance tasks, minimizing the logistical burden through resources required for maintenance and support, and reducing support costs [5]. Thus, some of the commonly addressed maintainability design factors are accessibility, test points, controls, labeling and coding, handles, standardization, lubrication, interchangeability, modular design, ease of removal and replacement, displays, skill requirements, indication and location of failures, and safety factors [5].

Over the years, many methods have been developed for performing various types of reliability and quality analyses. These methods include failure modes and effect analysis, fault tree analysis, Markov method, total quality management, and cause and effect diagram [4, 10, 13]. The first three of these methods are described in Sect. 3.5. The remaining two (*i.e.*, total quality management and cause-and-effect diagram) are presented below.

3.8.1 *Total Quality Management*

This method may simply be described as a philosophy of pursuing continuous improvement in all processes through the integrated or team efforts of all personnel in an organization. Over the years, the method has proven to be an effective tool to organizations in pursuit of improving the maintainability aspect of their products. Continuous improvement and customer satisfaction are the two fundamental principles of total quality management (TQM). In addition, seven important elements of TQM are team effort, supplier participation, management commitment and leadership, statistical tools, cost of quality, customer service, and training [27].

TQM can be implemented by following the five steps listed below [10]:

- **Step 1:** Create a vision.
- **Step 2:** Plan an appropriate action.
- **Step 3:** Create an effective structure (*e.g.*, eliminate roadblocks, involve employees, create cross-functional teams, and institute training).
- **Step 4:** Measure progress through appropriate means.
- **Step 5:** Update plans and vision.

In the past, many organizations have experienced various difficulties in the implementation of TQM. These difficulties include failure of senior management to delegate decision-making authority to lower organizational levels, failure of top management to devote sufficient time to the effort, and insufficient allocation of resources for training and developing manpower [28]. Additional information on TQM is available in Ref. [28].

3.8.2 Cause and Effect Diagram

This is a deductive analysis method developed by K. Ishikawa of Japan [29], and it can be quite useful in performing maintainability analysis. The method is also known as a fishbone diagram because of its resemblance to the skeleton of a fish. The right side (*i.e.*, the fish head) of the diagram denotes the effect (*e.g.*, the problem), and to the left of this are all possible problem-related causes connected to the central fish spine.

The cause and effect diagram can be developed by following the five steps listed below.

- **Step 1:** Identify the effect to be investigated or develop a problem statement.
- **Step 2:** Brainstorm to identify problem-related causes.
- **Step 3:** Group main causes into appropriate categories and stratify them.
- **Step 4:** Construct the diagram by linking the problem-related causes under appropriate process steps and write down the effect/problem in the diagram box (*i.e.*, the fish head) on the right side.
- **Step 5:** Refine cause categories by asking questions such as "What is the real reason of the existence of this condition?" and "What causes this?"

Some of the main benefits of the cause and effect diagram are its usefulness in generating ideas, in identifying root causes of the problem, in presenting an orderly arrangement of theories, and in guiding further inquiry.

Additional information on cause and effect diagrams is available in Ref. [29].

3.9 Maintainability-Management-Related Tasks During the Equipment Life Cycle

The life cycle of a piece of equipment may be divided into four phases: concept-development phase, validation phase, production phase, and operation phase. During its life cycle, to handle maintainability issues, various types of maintainability-management-related tasks are performed. In the concept-development phase, the maintainability management tasks are basically concerned with determining the equipment effectiveness needs, in addition to determining, from the equipment's purpose and intended operation, the required field support policies and other provisions.

In the validation phase, the maintainability management tasks include developing a maintainability program plan that satisfies contractual requirements, performing maintainability allocations and predictions, coordinating and monitoring maintainability efforts throughout the organization, participating in design reviews, developing a plan for maintainability testing and demonstration, and developing a planning document for data collection, analysis, and evaluation [30].

In the production phase, the maintainability management tasks include evaluating all proposals for changes with respect to their impact on maintainability, evaluating production test trends from the standpoint of adverse effects on maintainability requirements, monitoring production processes, etc. Finally, during the operation phase, although there are no specific maintainability management tasks, the phase is probably the most significant because during this period the equipment's true logistical support and cost effectiveness are demonstrated.

3.10 Need for Safety and Safety-Related Facts and Figures

Safety has become an important issue because each year a vast number of people die and get seriously injured due to various types of accidents. For example, in 1996 in the USA alone, according to the National Safety Council (NSC), there were 93,400 deaths and a large number of disabling injuries due to accidents [31]. Other factors that also play an important role in demanding the need for better safety include public pressures, government regulations, and the increasing number of lawsuits.

Some of the important safety-related facts and figures are as follows:

- In 2000, there were approximately 97,300 unintentional injury deaths in the USA and their cost to the US economy was estimated to be approx. $512.4 million [32].
- In a typical year in the USA, approximately 35 million work-hours are lost due to accidents [33].
- In the 1990s, the average annual cost of accidents per worker in the USA was approximately $420 [11].

- Over the 40-year period 1960–2000 work-related accidental deaths in the USA dropped by 60% [34, 35].
- In 1980, employers in the USA spent approximately $22 billion to insure or self-insure against work-related injuries [36].

Additional safety-related factors and figures are available in Ref. [13].

3.11 Equipment Hazard Classifications and Common Mechanical Injuries

There are many equipment-related hazards. They may be grouped under six distinct classifications: energy hazards, kinematic hazards, electrical hazards, misuse-and-abuse-related hazards, environmental hazards, and human-factor hazards [37]. All these classifications are described in detail in Ref. [37].

Humans interact with various types of equipment in the industrial sector to perform tasks such as drilling, chipping, shaping, cutting, punching, stitching, stamping, and abrading. Past experiences indicate that various types of injuries can occur in performing such tasks. The common ones include injuries related to shearing, crushing, puncturing, breaking, straining and spraining, and cutting and tearing. Additional information on all these injuries is available in Ref. [11].

3.12 Safety Analysis Methods

Over the years, a large number of methods have been developed to perform analysis in safety and related areas [4, 13]. These methods include failure modes and effect analysis, fault tree analysis, Markov method, cause and effect diagram, hazard and operability analysis, job safety analysis, and technic of operations review. The first four of these methods are described earlier in the chapter. The remaining three methods (*i.e.*, hazard and operability analysis, job safety analysis, and technic of operations review) are presented below.

3.12.1 Hazard and Operability Analysis (HAZOP)

This method was developed for application in the chemical industry, and its fundamental objectives are as follows [11, 13, 38–41]:

- To produce a complete description of a process or facility,
- To review each process/facility element for determining how deviations from the design intentions can occur, and
- To decide whether the deviations can result in operating problems or hazards.

A HAZOP study can be conducted by following the seven steps listed below [13,40].

- **Step 1:** Select the process system to be analyzed.
- **Step 2:** Form the team of experts.
- **Step 3:** Explain the HAZOP process to all team members.
- **Step 4:** Establish goals and appropriate time schedules.
- **Step 5:** Conduct brainstorming sessions as appropriate.
- **Step 6:** Perform analysis.
- **Step 7:** Document the study.

Additional information on HAZOP is available in Refs. [11,42].

3.12.2 Job Safety Analysis

This method is used to find and rectify potential hazards that are intrinsic to or inherent in a given workplace. Generally, people who participate in performing job safety analysis are worker, supervisor, and safety professionals. Job safety analysis can be performed by following the five steps listed below [43].

- **Step 1:** Choose the job to be analyzed.
- **Step 2:** Break down the job under consideration into a number of steps/tasks.
- **Step 3:** Identify potential hazards and propose necessary actions to control them to appropriate levels.
- **Step 4:** Apply the proposed actions.
- **Step 5:** Evaluate the results.

All in all, past experience indicates that the success of this method very much depends on the degree of rigor exercised by the job safety analysis team members throughout the analysis process.

3.12.3 Technic of Operations Review (TOR)

This is basically a hands-on analytical methodology used for determining the root system causes of an operation failure. The method or methodology was developed by D.A. Weaver of the American Society of Safety Engineers (ASSE) in the early 1970s [11]. It makes use of a worksheet containing simple terms requiring yes/no decisions. The basis for the activation of TOR is an incident occurring at a specific time and place involving certain individuals. The method is composed of the following steps [11,44]:

- **Step 1:** Form the TOR team with individuals having appropriate expertise and experience.
- **Step 2:** Hold a roundtable session with all members of the team.

- **Step 3:** Identify one important factor that has been pivotal in the occurrence of an incident or accident.
- **Step 4:** Use the team consensus in responding to a sequence of yes/no options.
- **Step 5:** Evaluate identified factors as well as ensure the existence of consensus among the team members.
- **Step 6:** Prioritize contributing factors.
- **Step 7:** Develop preventive/corrective strategies with respect to each contributing factor.
- **Step 8:** Implement strategies.

Additional information on TOR is available in Refs. [11, 44].

3.13 Safety Indexes

Over the years, many safety indexes have been developed to measure the safety performance of organizations. This section presents two of these safety indexes proposed by the American National Standards Institute [45].

3.13.1 Index I: Disabling Injury Frequency Rate

This index is defined by

$$DI_{\mathrm{fr}} = \frac{N(1{,}000{,}000)}{T_{\mathrm{e}}}, \tag{3.46}$$

where

DI_{fr} is the disabling injury frequency rate,
N is the number of disabling injuries,
T_{e} is the employee exposure time expressed in hours.

The index is based on four events (*i.e.*, deaths, permanent disabilities, permanent partial disabilities, and temporary disabilities) that occur during the period covered by the rate. Additional information on the index is available in Refs. [45, 46].

3.13.2 Index II: Disabling Injury Severity Rate

This index is defined by

$$DI_{\mathrm{sr}} = \frac{d_{\mathrm{c}}(1{,}000{,}000)}{T_{\mathrm{e}}}, \tag{3.47}$$

where

DI_{sr} is the disabling injury severity rate,
d_{c} is the number of days charged.

The index is based on four factors [*i.e.*, total scheduled charges (days) for all deaths, permanent total disabilities, permanent partial disabilities, and the number of days of disability from all temporary injuries] occurring during the time period covered by the rate. Additional information on the index is available in Refs. [45, 46].

3.14 Problems

1. Discuss the bathtub hazard rate curve.
2. Prove Eqs. (3.7)–(3.9) using Eq. (3.3).
3. Prove Eq. (3.22).
4. Describe failure modes and effect analysis.
5. Prove that the sum of Eqs. (3.37) and (3.38) is equal to unity.
6. Prove Eq. (3.38) using Eqs. (3.35) and (3.36).
7. Define the following terms:
   - OR gate
   - AND gate
8. Define maintainability function.
9. Describe total quality management.
10. What are the common equipment-related injuries?

References

1. Layman, W.J.: Fundamental consideration in preparing a master system plan. Electric. World 101, 778–792 (1933)
2. Smith, S.A.: Service reliability measured by probabilities of outage. Electric. World 103, 371–374 (1934)
3. Dhillon, B.S.: Power System Reliability, Safety, and Management. Ann Arbor Science Publishers, Ann Arbor, MI (1983)
4. Dhillon, B.S.: Design Reliability: Fundamentals and Applications. CRC, Boca Raton, FL (1999)
5. AMCP706-133, Engineering Design Handbook: Maintainability Engineering Theory and Practice, US Department of Defense, Washington, DC (1976)
6. Akenbrandt, F.L. (ed.): Electronic Maintainability. Engineering Publishers, Elizabeth, NJ (1960)
7. MIL-STD-470, Maintainability Program Requirements, US Department of Defense, Washington, DC (1966)
8. MIL-HDBK-472, Maintainability Prediction, US Department of Defense, Washington, DC (1966)
9. Retterer, B.L., Kowalski, R.A.: Maintainability: a historical perspective. IEEE Trans. Reliabil. 33, 56–61 (1984)
10. Dhillon, B.S.: Engineering Maintainability. Gulf, Houston, TX (1999)
11. Goetsch, D.L.: Occupational Safety and Health. Prentice-Hall, Englewood Cliffs, NJ (1996)
12. Heinrich, H.W.: Industrial Accident Prevention, 3rd edn. McGraw-Hill, New York (1950)

13. Dhillon, B.S.: Engineering Safety: Fundamentals, Techniques, and Applications. World Scientific, River Edge, NJ (2003)
14. Kapur, K.C.: Reliability and maintainability. In: Salvendy, G. (ed.) Handbook of Industrial Engineering. Wiley, New York, pp. 8.5.1–8.5.34 (1982)
15. Dhillon, B.S.: Mechanical Reliability: Theory, Models, and Applications, American Institute of Aeronautics and Astronautics, Washington, DC (1988)
16. Lipp, J.P.: Topology of Switching Elements vs. Reliability. IRE Reliabil. Qual. Control 7, 21–34 (1957)
17. Elsayed, E.A.: Reliability Engineering. Addison Wesley Longman, Reading, MA (1996)
18. Ramakumar, R.: Engineering Reliability: Fundamentals and Applications. Prentice-Hall, Englewood Cliffs, NJ (1993)
19. Dhillon, B.S.: Advanced Design Concepts for Engineers. Technomic, Lancaster, PA (1998)
20. Countinho, J.S.: Failure effect analysis. Trans. N.Y. Acad. Sci. 26, 564–584 (1964)
21. Shooman, M.L.: Probabilistic Reliability: An Engineering Approach. McGraw-Hill, New York (1968)
22. Vesley, W.E., Goldberg, F.F., Roberts, N.H., Haasal, D.F.: Fault Tree Handbook, Report No. NUREG-0492, US Nuclear Regulatory Commission, Washington, DC (1981)
23. Latino, C.J.: Hidden Treasure: Eliminating Chronic Failures Can Cut Maintenance Costs Up to 60%, Report, Reliability Center, Hopewell, VA (1999)
24. AMCP-706-134, Engineering Design Handbook: Maintainability Guide for Design, US Department of Defense, Washington, DC (1972)
25. Blanchard, B.S., Verma, D., Peterson, E.L.: Maintainability. Wiley, New York (1995)
26. Von Alven, W.H. (ed.): Reliability Engineering. Prentice-Hall, Englewood Cliffs, NJ (1964)
27. Burati, J.L., Matthews, M.F., Kalidindi, S.N.: Quality management organizations and techniques. J. Construct. Eng. Manage. 118, 112–128 (1992)
28. Gevirtz, C.D.: Developing New Products with TQM. McGraw-Hill, New York (1994)
29. Ishikawa, K.: Guide to Quality Control. Asian Productivity Organization, Tokyo, 1976.
30. Dhillon, B.S., Reiche, H.: Reliability and Maintainability Management. Van Nostrand Reinhold, New York (1985)
31. Accident Facts, Report, National Safety Council, Chicago (1996)
32. Report on Injuries in America 2000, National Safety Council, Chicago (2000)
33. Accident Facts, National Safety Council, Chicago (1990–1993)
34. Report on Injuries in America, National Safety Council, Chicago (2001)
35. Blake, R.P. (ed.): Industrial Safety. Prentice-Hall, Englewood Cliffs, NJ (1964)
36. Lancianese, F.: The soaring costs of industrial accidents. Occupat. Hazards pp. 30–35 (August 1983)
37. Hunter, T.A.: Engineering Design for Safety. McGraw-Hill, New York (1992)
38. Guidelines for Hazard Evaluation Procedures, American Institute of Chemical Engineers, New York (1985)
39. Dhillon, B.S., Rayapati, S.N.: Chemical systems reliability: a survey. IEEE Trans. Reliabil. 37, 199–208 (1988)
40. Roland, H.E., Moriarty, B.: System Safety Engineering and Management. Wiley, New York (1983)
41. Gloss, D.S., Wardle, M.G.: Introduction to Safety Engineering. Wiley, New York (1984)
42. Risk Analysis Requirements and Guidelines, Report No. CAN/CSA-Q634-91, prepared by the Canadian Standards Association, 1991. Available from the Canadian Standards Association, 178 Rexdale Boulevard, Rexdale, Ontario, Canada
43. Hammer, W., Price, D.: Occupational Safety Management and Engineering. Prentice-Hall, Upper Saddle River, NJ (2001)
44. Hallock, R.G.: Technic of operations review analysis: determine cause of accident/incident. Safety Health 60, 38–39, 46 (August 1991)
45. Z-16.1, Method of Recording and Measuring Work Injury Experience, American National Standards Institute, New York (1985)
46. Tarrants, W.E.: The Measurement of Safety Performance. Garland STPM, New York (1980)

Chapter 4
Mining Equipment Reliability

4.1 Introduction

Each year billions of dollars is spent to produce various types of equipment for use by the mining industry throughout the world, and this expenditure is increasing rapidly. For example, as per the United States Census Bureau, in 2004 American mining equipment manufacturers shipped around $1.4 billion worth of goods, and a year later that figure jumped to $2 billion [1]. Today, the global economy is forcing mining companies to modernize their operations through increased mechanization and automation.

Thus, as mining equipment is becoming more complex and sophisticated, its cost is increasing rapidly. This in turn makes it cost ineffective to have standby units. To meet production targets, mining companies are increasingly demanding better equipment reliability. Reliability is a good performance indicator of overall equipment condition and is defined as the probability that a piece of equipment will perform its function satisfactorily for the desired period of time when used according to specified conditions. However, in the industrial sector reliability is frequently expressed in terms of mean time between failures.

This chapter presents various important aspects of mining equipment reliability.

4.2 Reasons for Improving Mining Equipment Reliability, Factors Impacting Mining System Reliability, and Useful Mining-Equipment-Reliability-Related Measures

There are many reasons for improving mining equipment reliability. Some of these are as follows [2]:

- To maximize profit.
- To reduce the cost of poor reliability (the true cost of poor reliability in most mining operations, when measured effectively, is quite significant).

- To reduce the performance of mining equipment services in an unplanned manner because of short notice.
- To provide more accurate short-term forecasts for equipment operating hours.
- To overcome challenges imposed by global competition.
- To take advantage of lessons learned from other industrial sectors such as aerospace, defense, and nuclear power generation.

There are many factors that, directly or indirectly, impact mining system reliability. Most of these factors are shown in Fig. 4.1 [2].

In the case of geology, variability in digging conditions can result in the need for trucks or shovels to stop. Similarly, different ore grades than the one expected can lead to the need to interrupt the production process. In the case of weather, rain or fog can interrupt production.

In the case of the mine plan, it normally calls for equipment to be shifted on a periodic basis as different areas are to be mined. In turn, this leads to an interruption to a steady-state production. In the case of the blast, there is frequently a need to stop the equipment operation. In the case of accident damage, it causes an interruption to the ongoing production process if the equipment has to be taken out of service for inspection or repair.

In the case of refueling and lubrication, the stoppage of equipment for refueling and lubrication results in the interruption of the production process. In the case of equipment failure, it is basically a maintenance issue and it causes an interruption

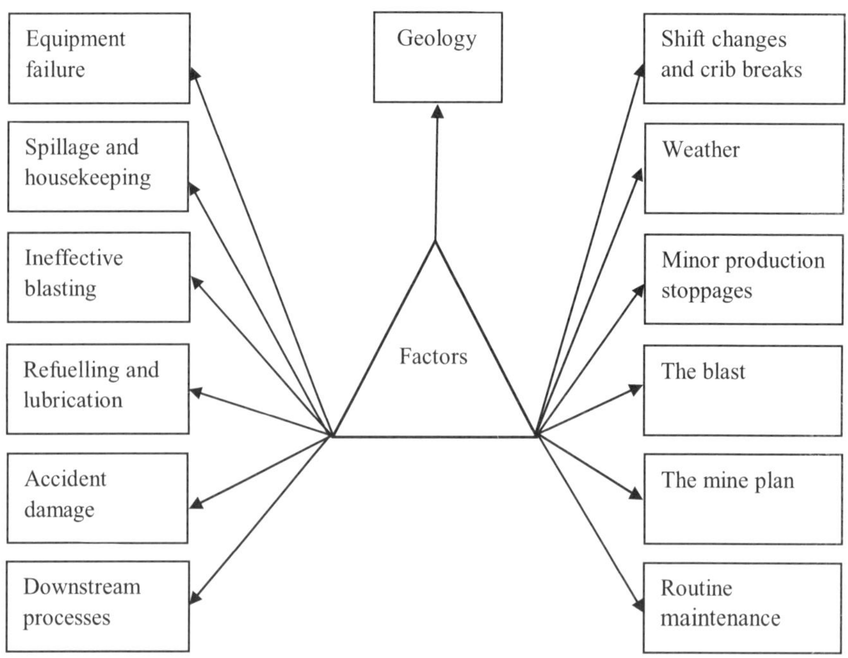

Fig. 4.1 Factors directly or indirectly impacting mining system reliability

in the production process. In the case of spillage and housekeeping, the need to stop and clean up spillage around the shovel or the dump area causes an interruption in the production process as well.

In the case of routine maintenance, routine servicing, overhauls, and component replacements lead to interruptions in production while the equipment is taken out of service. In the case of ineffective blasting, it can lead to problems such as poor diggability in certain areas and unreliable equipment operation. In the case of downstream processes, if a downstream process stops in a direct tipping situation, this can lead to an interruption in the mining operation.

In the case of minor production stoppages, "comfort stops" such as minor adjustments can interrupt the production process. Finally, in the case of shift changes and crib breaks, every shift change and crib break usually leads to an interruption to the steady-state nature of the production operation.

4.2.1 Useful Mining-Equipment-Reliability-Related Measures

There are many mining-equipment-reliability-related measures. Some of these are presented below [2].

Availability

This is simply the proportion of time the equipment is able to be used for its intended purpose and is expressed by

$$AV_{\mathrm{m}} = \frac{(TH - DT)}{TH}(100)\,, \tag{4.1}$$

where

AV_{m} is the mining equipment availability,
TH is the total hours,
DT is the downtime expressed in hours.

Mean Time Between Failures

This is expressed by

$$MTBF = \frac{TH - DT - SH}{NF}\,, \tag{4.2}$$

where

$MTBF$ is the mean time between failures,
SH is the standby hours,
NF is the number of failures.

Utilization

This is defined by

$$U = \frac{(TH - DT - SH)(100)}{(TH - DT)}, \quad (4.3)$$

where

U is the utilization.

Production Efficiency

This may be described simply as the ratio of actual output from a piece of equipment/machine (which satisfies the required quality standards) to its rated output during the period it is operational. Nonetheless, production efficiency is expressed by

$$PE = \left[\left\{\frac{AP}{TH - DT - SH}\right\} / RC\right](100), \quad (4.4)$$

where

PE is the production efficiency,
AP is the actual production,
RC is the rated capacity expressed in units per hour.

Overall Equipment Effectiveness

This is expressed by

$$OEE = (AV_{\mathrm{m}})(PE)(U), \quad (4.5)$$

where

OEE is the overall equipment effectiveness.

4.3 Open-Pit-System Reliability Analysis

The selection of equipment for modern open-pit mines has become a challenging issue in terms of productivity, reliability, availability, maintainability, etc. The overall system has grown to the level where the application of reliability principles has been conceived to be quite useful for meeting the ever-growing technological requirements. The system is composed of various types of loading and dumping machinery that, in turn, are arranged in different arrays. These arrays result in various types of sequencing systems. The failure of a single element in the sequences can cause part or complete system failure.

Nonetheless, each element of open-pit mines (*i.e.*, working face, shovel, dumper, dumping point, etc.) may be considered as an independent link to the open-pit mine chain system. Reliability analyses of the chain system when it forms series and parallel configurations are presented below [3].

4.3.1 Open-Pit Series System

In this case, the open-pit-mine components, *i.e.*, working face, shovel, dumper, and pumping place, form a series configuration. This means all system components must work normally for the system to succeed.

For independent components, the system reliability is given by [3,4]

$$R_{ops} = R_s R_d R_{wf} R_{dp} , \tag{4.6}$$

where

R_{ops}	is the open-pit series system reliability,
R_s	is the shovel reliability,
R_d	is the dumper or dump-truck reliability,
R_{wf}	is the working-face reliability,
R_{dp}	is the dumping-place reliability.

For constant failure rates of the shovel, dumper or dump truck, working face, and dumping place, using Chap. 3 and Ref. [4] we get

$$R_s(t) = e^{-\lambda_s t} , \tag{4.7}$$

$$R_d(t) = e^{-\lambda_d t} , \tag{4.8}$$

$$R_{wf}(t) = e^{-\lambda_{wf} t} , \tag{4.9}$$

$$R_{dp}(t) = e^{-\lambda_{dp} t} , \tag{4.10}$$

where

$R_s(t)$	is the shovel reliability at time t,
$R_d(t)$	is the dumper or dump-truck reliability at time t,
$R_{wf}(t)$	is the working-face reliability at time t,
$R_{dp}(t)$	is the dumping-place reliability at time t,
λ_s	is the shovel constant failure rate,
λ_{wf}	is the working face constant failure rate,
λ_d	is the dumper or dump-truck constant failure rate,
λ_{dp}	is the dumping-place constant failure rate.

Substituting Eqs. (4.7)–(4.10) into Eq. (4.6) we get

$$R_{ops}(t) = e^{-\lambda_s t} \cdot e^{-\lambda_d t} \cdot e^{-\lambda_{wf} t} \cdot e^{-\lambda_{dp} t} = e^{-(\lambda_s + \lambda_d + \lambda_{wf} + \lambda_{dp})t} , \tag{4.11}$$

where

$R_{ops}(t)$	is the open-pit series system reliability at time t.

Integrating Eq. (4.11) over the time interval $[0,\infty]$, we obtain the following expression of the open-pit series system mean time to failure [4]:

$$MTTF_{\text{ops}} = \int_0^\infty R_{\text{ops}}(t)\,\mathrm{d}t = \frac{1}{\lambda_s + \lambda_d + \lambda_{wf} + \lambda_{dp}}, \tag{4.12}$$

where

$MTTF_{\text{ops}}$ is the open-pit series system mean time to failure.

Example 4.1

Assume that an open-pit system is composed of shovel, dump truck, working face, and dumping place, which form a series configuration. The reliabilities of the four components are 0.8, 0.9, 0.98, and 0.97, respectively. Calculate the open-pit system reliability if all its components fail independently.

Substituting the specified data values into Eq. (4.6) we get

$$R_{\text{ops}} = (0.8)(0.9)(0.98)(0.97) = 0.6844\,.$$

Thus, the open-pit system reliability is 0.6844.

Example 4.2

Assume that in Example 4.1, the constant failure rates of shovel, dump truck, working face, and dumping place are 0.005 failures per hour, 0.006 failures per hour, 0.007 failures per hour, and 0.008 failures per hour, respectively. Calculate the open-pit series system mean time to failure.

Using the above given data values in Eq. (4.12) yields

$$MTTF_{\text{ops}} = \frac{1}{0.005 + 0.006 + 0.007 + 0.008} = 38.46\,\text{h}\,.$$

Thus, the mean time to failure of the open-pit series system is 38.46 h.

4.3.2 Open-Pit Parallel System

This type of configuration is formed when there is more than one unit of an open-pit system's components operating simultaneously and at least one of these units must work normally for the system to succeed. For example, there are two shovels operating simultaneously and at least one of the shovels must work normally for the system to succeed. In this case, both shovels form a parallel configuration and the configuration's reliability, if both shovels fail independently, using Chap. 3 and Ref. [4], is expressed by

$$R_{ps} = 1 - (1 - R_{S1})(1 - R_{S2}) , \tag{4.13}$$

where

R_{ps} is the shovel parallel configuration reliability,
R_{S1} is the reliability of shovel number one,
R_{S2} is the reliability of shovel number two.

For constant failure rates of the shovels, using Chap. 3 and Ref. [4] we write

$$R_{S1}(t) = e^{-\lambda_{S1} t} , \tag{4.14}$$

$$R_{S2}(t) = e^{-\lambda_{S2} t} , \tag{4.15}$$

where

$R_{S1}(t)$ is the reliability of shovel number one at time t,
$R_{S2}(t)$ is the reliability of shovel number two at time t,
λ_{S1} is the constant failure rate of shovel number one,
λ_{S2} is the constant failure rate of shovel number two.

Inserting Eqs. (4.14) and (4.15) into Eq. (4.13) we get

$$R_{ps}(t) = 1 - \left(1 - e^{-\lambda_{S1} t}\right)\left(1 - e^{-\lambda_{S2} t}\right) = e^{-\lambda_{S1} t} + e^{-\lambda_{S2} t} - e^{-(\lambda_{S1} + \lambda_{S2}) t} , \tag{4.16}$$

where

$R_{ps}(t)$ is the shovel parallel configuration reliability at time t.

Integrating Eq. (4.16) over the time interval $[0, \infty]$, we obtain the following expression for the shovel parallel configuration mean time to failure:

$$MTTF_{ps} = \int_0^\infty R_{ps}(t)\,dt = \frac{1}{\lambda_{S1}} + \frac{1}{\lambda_{S2}} - \frac{1}{(\lambda_{S1} + \lambda_{S2})} , \tag{4.17}$$

where

$MTTF_{ps}$ is the shovel parallel configuration mean time to failure.

Example 4.3

An open-pit system has two independent and identical shovels forming a parallel configuration (*i.e.*, at least one shovel must work normally for the system to succeed). The shovel constant failure rate is 0.008 failures per hour. Calculate the shovel parallel configuration reliability for a 200-h mission.

Substituting the specified data values into Eq. (4.16) we get

$$R_{ps}(200) = e^{-(0.008)(200)} + e^{-(0.008)(200)} - e^{-(0.008+0.008)(200)} = 0.3630\,.$$

Thus, the shovel parallel configuration reliability is 0.3630.

4.4 Programmable Electronic Mining System Failures

Various types of hazards can occur with programmable electronic mining system hardware or software failures [5]. Hardware failures are physical failures and are usually the result of wear and random events. They can involve any physical component of the system including sensors, data communication paths, power supplies, and programmable electronic devices. Random hardware failures include items such as broken wires, mechanical defects, corroded contacts, dielectric failures, open circuits, and short circuits.

In contrast, software failures occur due to systematic (functional) errors. Systematic errors include items such as software bugs, requirement errors, timing errors, design errors, operator errors, and management-of-change errors.

A study of data from the US Mine Safety and Health Administration (MSHA), New South Wales Mines in Australia, and Queensland Mines in Australia, concerning programmable electronic-based mining systems, revealed that there were a total of 100 mishaps during the period 1995–2001. *The* breakdown of these mishaps was as follows [5]:

- Random hardware failures: 46%
- Systematic failures: 39%
- Miscellaneous failures: 15%

Both random hardware failures and systematic failures are described below, separately.

4.4.1 Random Hardware Failures

The harsh environmental factors in mines, such as vibrations, heat, water and dirt intrusion, and shock, can greatly influence the occurrence of programmable electronic-based mining system hardware failures. These failures involve items such as sensors, wiring, electrical connectors, solenoids, and power supplies. An example of such failures is the degradation of rubber boots/seals used to keep out moisture and dust.

An analysis of the random hardware failures of the above data (*i.e.*, for programmable electronic-based mining systems for the period 1995–2001) revealed the following breakdown of the failures [5]:

- Sensor failures: 33%
- Electronic-component failures: 26%
- Moisture-related failures: 17%
- Actuator failures: 13%
- Miscellaneous failures: 11%

Note that in the above breakdown switch-related failures were included under the sensor failure category.

4.4.2 Systematic Failures

Periodic systematic failures are also called functional failures. Some of the sources of these failures are hardware and software design errors, errors resulting from software modifications, operator errors, and errors made during maintenance and repair activities. Analysis of systematic failures of the data for programmable electronic-based mining systems for the period 1995–2001 revealed the following breakdown of the failures [5]:

- Design-error-related failures: 50%
- Maintenance- and repair-error-related failures: 40%
- Miscellaneous failures: 10%

4.5 Designing Reliable Conveyor Belt Systems and Methods of Measuring Winder Rope Degradation

Today modern mining approaches have put greater emphasis than ever before on the reliability of the belt conveyor system that removes mined material, say coal, from the face. More specifically, with the application of modern long-wall methods, it is quite possible that up to 90% of the mine production comes from one face, and it, in turn, must be handled effectively by a single-gate conveyor. Thus, modern mines are looking for a conveyor reliability of close to 100%, and the conveyor industry is under increasing pressure to achieve this target.

This target or objective can be achieved by the industry by following the five design-related guidelines shown in Fig. 4.2 [6].

In underground mines, winders that use steel wire ropes are normally used for moving materials. During usage, the steel wire ropes are subjected to continuous degradation or deterioration. Therefore, the reliability of ropes is very important for the performance of shaft-equipped mines as well as for the safety of miners. Past experience indicates that the type of damage that occurs in winder ropes during their usage period includes, but is not limited to, wire fatigue and resulting breaks, abrasion formation of kinks and loops, and corrosion [7–9].

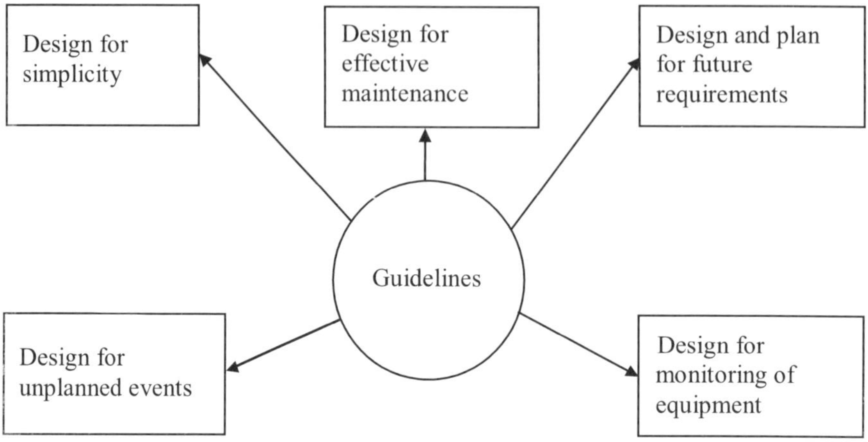

Fig. 4.2 Useful design-related guidelines for improving conveyor reliability

Due to safety concerns many mining regulatory authorities around the world mandate periodic inspections to be performed to determine the condition of winder ropes. Thus usually two types of inspections (*i.e.*, visual inspection and magnetic nondestructive testing) are performed as it is impossible to find rope internal damage and corrosion through visual inspection alone.

Both visual inspection and magnetic nondestructive testing methods are described below, separately.

4.5.1 Visual Inspection Method

This method is quite useful for identifying changes in rope diameter and lay length, wear of the crown wires, external wire breaks or loose wires, visible corrosion, and any other external damage. In fact, visual inspection is the only effective approach to identifying the severity of rope external abrasive wear as the magnetic approach tends to underestimate the crown wire wear.

Some of the steps of the visual inspection method are as follows [7, 10]:

(i) Measure the rope diameter and the lay length at a number of points or sections.
(ii) Examine for broken wires and excessive crown wear.
(iii) Examine the whole rope end to end for damage or abuse.
(iv) Examine the rope termination for corrosion, broken wires, and condition of fastening.
(v) Examine the sheaves for wear, misfit, and so on.
(vi) Examine the condition of the drum in the case of drum winders.
(vii) Check for appropriate lubrication.

4.5.2 Nondestructive Testing Method

Electromagnetic or permanent magnet-based methods are quite effective at detecting damage anywhere within a rope in both interior and exterior wires [7, 11, 12]. Magnetic rope testing is performed by passing the rope through a permanent magnet-based device or instrument. In this case, a length of the rope is fully magnetized as it passes through the test device. Magnetic rope testing devices are used for monitoring two distinct types (*i.e.*, types I and II) of magnetic field changes caused by the existence of anomalies.

Type I is known as loss of metallic area (LMA). It may be described simply as a relative measure of the amount of material mass missing from a cross-section in a wire rope. LMA is measured by comparing a section's magnetic field intensity with that of a reference section on the rope that denotes the maximum, unworn metallic cross-section area. Type II changes involve a magnetic dipole generated by a discontinuity in a magnetized section of the rope such as a wire break, a corrosion pit, or a groove worn into a wire. Often these are known as leakage flux (LF) flaws.

4.6 Fault Tree Analysis of Shovel Machine

Shovel machines are an important element of mining machinery. Their failures can have an important impact on the overall performance of mines with respect to production. Fault tree analysis, described in Chap. 3 and in Ref. [4], is an effective tool for performing failure analysis of shovel machines. A fault tree of a shovel machine is shown in Fig. 4.3 [13]. In this case, the shovel machine is divided into six subsystems: engine, bucket, transmission, hydraulic, track, and body and cabin. Failure of any one of these subsystems will cause the failure of the shovel machine. In turn, an engine failure can occur either due to a failure of either the static parts or the dynamic parts.

Failure in the static parts can be initiated either by engine block failure or by cylinder head failure. Similarly, the failure in the dynamic parts can be caused either by piston failure, crank failure, shaft failure, or bearing failure.

Hydraulic failure can occur either due to steering failure or bucket hydraulic failure. In turn, the occurrence of any of the following four failures can cause failure of the bucket hydraulic [12]:

- Tube failure
- Filter failure
- Pump failure
- Tank failure

Pump failure can be caused by the failure of pistons, seals, or valves.

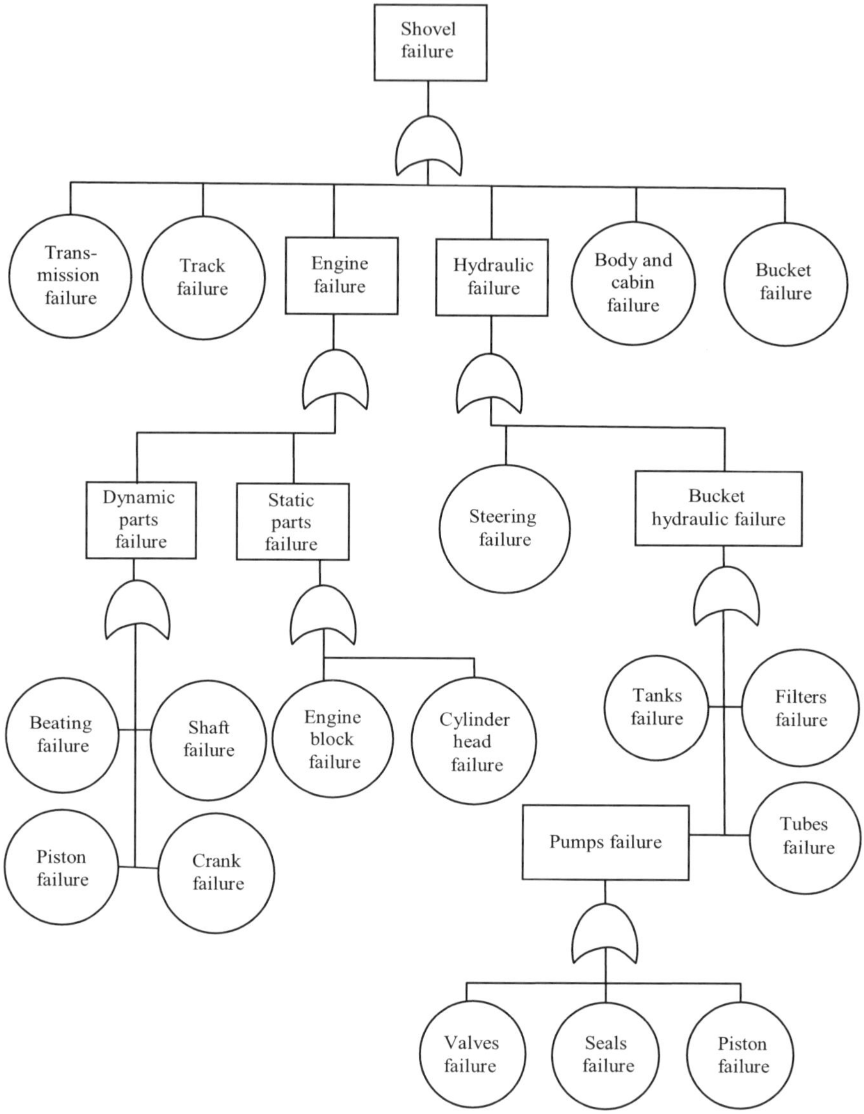

Fig. 4.3 Fault tree of a shovel machine

By examining Fig. 4.3, it can be concluded that the occurrence of any fault event shown in circles will cause shovel failure. The probability of occurrence of shovel failure can be calculated for known probabilities of occurrence of the fault events shown in circles, with the aid of Ref. [4].

4.7 Dump-truck Tire Reliability and the Factors Affecting Their Life

The shovel-truck system is probably the most flexible system in opencast mines. Its reliability and availability are the most important factors in successfully meeting the production target of opencast mines. Tires of dump trucks are an important element of the shovel-truck system. Their reliability directly or indirectly affects the overall production performance of opencast mines.

A study of dump-truck tires revealed that the tires' times to failure followed a normal distribution [14]. This simply means that their failure rate is nonconstant and the reliability of the tires must be computed by using the normal distribution representing the tire failure times.

There are many factors that can affect dump-truck tire life including underinflation, overinflation, tire bleeding, heat generation, and speed and haul length [14,15]. Underinflation can lead to excessive flexing of the sidewalls and an increase in internal tire temperature. In turn, this can cause permanent damage to tires such as radial cracks, ply separation, and casing breakup.

Overinflation reduces the amount of tread in contact with the ground and in addition to makes tires vulnerable to factors such as cuts, snags, and impact fractures. In the case of tire bleeding, lowering tire pressure after a long run is a normal practice. This increases tire temperature, causes premature failures, and makes an underinflated tire withstand high load (*i.e.*, vehicle weight).

In the case of heat generation, as rubber is a poor conductor, the heat generated through flexing will lead to heat accumulation. Note that, although the recommended load-speed inflation pressure ensures an equilibrium between heat generated and heat dissipated, any deviation can result in high temperature.

Finally, in the case of speed and haul length, speeds above 30 km per hour and a haul length of more than 5 km can considerably affect the life of tires.

4.8 Problems

1. What are the main reasons for improving mining equipment reliability?
2. List at least ten important factors that directly or indirectly impact mining system reliability.
3. List and define at least three mining-equipment-reliability-related measures.
4. Assume that an open-pit system is composed of shovel, dump truck, working face, and dumping place, which form a series of configurations. The constant failure rates of the four components are 0.0004 failures/h, 0.0005 failures/h, 0.0006 failures/h, and 0.0007 failures/h, respectively. Calculate the open-pit system reliability for a 100-h mission if all its components fail independently.
5. Discuss programmable electronic mining system failures.
6. Discuss useful design-related guidelines for improving conveyor reliability.

7. Discuss the visual inspection method with regard to inspecting winder ropes.
8. What is a shovel machine?
9. Write an essay on mining equipment reliability.
10. Discuss the factors that can affect the life of dump-truck tires.

References

1. Chadwick, J., Higgins, S.: US technology. Int. Min. pp. 44–54 (September 2006)
2. Dunn, S.: Optimising production scheduling for maximum plant utilisation and minimum downtime: the reliability revolution. Presented at the Dollar Driven Mining Conference, Perth, Australia, July 1997. Available online at http://www.plantmaintenance.com/ops.shtml
3. Mukhopadhyay, A.K.: Open pit system reliability. J. Mines Metals Fuels, 389–392 (August 1988)
4. Dhillon, B.S., Design Reliability: Fundamentals and Applications. CRC, Boca Raton, FL (1999)
5. Sammarco, J.J.: Programmable Electronic Mining Systems: Best Practices Recommendations (in Nine Parts), Report No. IC 9480, (Part 6: 5.1 System Safety Guidance), National Institute for Occupational Safety and Health (NIOSH), US Department of Health and Human Services, Washington, DC (2005). Available from the NIOSH: Publications Dissemination, 4676 Columbia Parkway, Cincinnati, OH 45226-1998
6. Moody, C.: Reliable Conveyor Belt Design, Proceedings of the American Mining Congress, pp. 579–582 (1991)
7. Kuruppu, M.: Methods and Reliability of Measuring Winder Rope Degradation. Mine Plann. Equip. Select. 261–266 (April 2003)
8. Chaplin, C.R.: Hoisting ropes for drum winders: the mechanics of degradation. Min. Technol. 213–219 (1994)
9. Chaplin, C.R.: Failure mechanisms in wire ropes. Eng. Failure Anal. 2, 45–57 (1995)
10. ASTM E157-93, Standard Practice for Electromagnetic Examination of Ferro-Magnetic Steel Wire Ropes, American Society for Testing and Materials (ASTM), Philadelphia (1993)
11. Weischedal, H.R.: The inspection of wire ropes in service: a critical review. Material Evaluation, 43, 13, 1592–1605 (1985)
12. Poffenroth, D.N.: Procedures and Results of Electromagnetic Testing of Mine Hoist Ropes Using the LMA-TEST Instruments, Proceedings of the OIPEEC Round Table Conference, pp. 17–21 (September 1989)
13. Samanta, B., Sarkar, B., Mukherjee, S.K.: Reliability centred maintenance (RCM) for heavy earth-moving machinery in an open cast coal mine. CIM Bull. 104–108 (November/December 2001)
14. Dey, A., Bhattacharya, J., Banerjee, S.: Prediction of field reliability for dumper tyres. Int. J. Surf. Min. Reclamat. Environ. 8, 23–25 (1994)
15. Balen, O.: Off the road tyres: correct selection and proper maintenance. J. Mines Metal Fuels 22, 107–113 (1979)

Chapter 5
Human Factors and Error in Mining

5.1 Introduction

The field of human factors exists because people make various types of errors in using engineering systems. Human factors may simply be described as a body of scientific facts concerning the characteristics of human beings [1–3]. Similarly, human error is defined as the failure to perform a specified task (or the performance of a forbidden action) that could result in the disruption of scheduled operations or damage to equipment and property [2, 3].

The history of the study of human factors may be traced back to 1898, when Frederick W. Taylor performed various studies to determine the most effective design of shovels [4]. By 1945 human-factor engineering was recognized as a specialized discipline, and in 1958 H.L. Williams recognized that human-element reliability must be taken into consideration in the overall system-reliability prediction; otherwise, such a prediction would not be realistic [5]. In 1960, a study conducted by the US military reported that human error is the cause of 20 to 50% of equipment failures [6]. Today, a wide range of published literature is available on various aspects of human factors including human error [3, 7, 8].

In the area of human factors in mining, the first formal human-factor study was conducted in 1971 and was concerned with identifying human-factor-related problems in underground coal mines [9, 10]. In 1982, two human-factor-problem-identification studies directed at surface mining were reported: one concerning the mining process itself and the other the processing plants [11, 12].

Over the years many publications on various aspects of human factors in mining including human error have appeared [10, 13]. This chapter presents various important aspects of human factors and error that are, directly or indirectly, concerned with mining.

5.2 The Need to Apply Human Factors in Mining and Common Roadblocks to the Introduction of Human Factors in an Organization

Technology plays a pivotal role in increasing productivity and safety in the mining industry. New advances in technology have placed new demands on mine workers. This, in turn, has created the need for paying greater attention to determine how people and technology can work together effectively. More specifically, how should mining equipment and tasks be designed to match the capabilities and limitations of mine workers who will be operating and maintaining such equipment in the field [10]?

The application of human-factor principles can be very useful in addressing questions such as this.

Over the years various types of roadblocks to the application of human factors in organizations have been experienced. Some of the common ones are shown in Fig. 5.1 [10]. Counterarguments to each of these roadblocks are presented below.

To the notion that all humans are created equal can be added that no two humans are identical. Thus, systems must be designed in such a way that they effectively accommodate the diversity in the human population. As for the idea that humans can be trained to overcome design-related shortcomings, although this is true to a certain degree, training can be somewhat unreliable. More specifically, as experience has shown, under stressful conditions humans usually respond the way they think systems/equipment should operate, which may not be how the systems/equipment actually operate.

Regarding the idea that engineers and designers know best how humans think and act, one may add that engineers and designers may know how engineers and design-

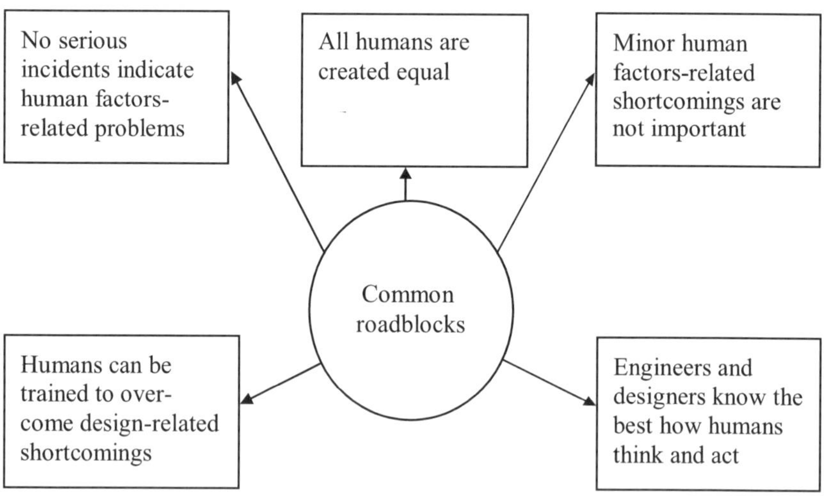

Fig. 5.1 Common roadblocks to the application of human factors in organizations

ers usually think and act, but they (*i.e.*, engineers and designers) do not necessarily think or act like other members of the society at large.

In regard to the idea that no serious incidents indicate human-factor-related problems, one may add that, although the Three Mile Island Nuclear Generating Station had various human-factor-related shortcomings in the control room, until the very day it came within 60 min of meltdown, the station reported no serious incidents [10].

Finally, the idea that minor human-factor-related shortcomings are not important is undermined by past experiences, which shows that minor shortcomings often compound into major shortcomings and usually have a way of insidiously eating into efficiency and productivity.

5.3 Human Sensory Capacities and Human-Factor Considerations in Equipment Design

Human sensory capacities include touch, sight, noise, vibration, and smell. Humans can recognize a minute change in these sensors over a wide range and then react to them automatically. As in the area of mining equipment, human sensory capabilities play an important role, and four of the five above-mentioned capacities are discussed below, separately [14–16].

Sight

Sight is stimulated by electromagnetic radiation of certain wavelengths known as the visible segment of the electromagnetic spectrum. Various parts of this spectrum, as seen by the human eye, appear to vary quite significantly in brightness. For example, in the daylight, the eye is quite sensitive to greenish-yellow light, and from different angles it sees differently.

Although looking straight ahead human eyes can perceive all colors, with an increase in the viewing angle color perception begins to decrease.

Some of the useful sight-related guidelines for designers are as follows [15, 16]:

- Choose colors such that color-weak individuals do not get confused.
- Do not rely too much on color when critical operations or tasks are to be performed by fatigued personnel.
- Try to use red filters with a wavelength greater than 6500 Angstrom units.

Vibration

Vibration is basically concerned with the effects on the performance of humans of periodically occurring mechanical forces impinging on body tissues. Past experience shows that the unsatisfactory performance of mental and physical tasks by people such as equipment operators and maintainers could be partially or wholly due to vibrations. For example, low-frequency and large-amplitude vibrations can lead to headaches, motion sickness, fatigue, deterioration in the ability to read and interpret instruments, and eye strain.

Some of the useful guidelines to reduce the effects of vibration and motion are as follows [15–18]:

- Use cushioned seats or dumping material seats to reduce vibration transmissions.
- Eliminate vibrations with an amplitude greater than 0.08 mm for critical maintenance or other tasks requiring letter or digit discrimination.
- Use devices such as springs, shock absorbers, and cushion mountings wherever possible.

Touch

Touch is closely associated with humans' ability to interpret auditory and visual stimuli. It adds, or sometime may even replace, the information transmitted to the brain by the ears and eyes. Thus, in mining equipment design the touch sensor can be utilized to relieve ears and eyes of a part of the load.

An example of the touch sensor's application could be the recognition of control knob shapes with or without the use of other sensors.

Noise

Noise may be described as sounds that lack coherence; it can affect the performance quality of a task requiring intense concentration. Past experience shows that noise contributes to human feelings such as well-being, irritability, and boredom.

Although the ear can detect sounds of frequencies from 20 to 20,000 Hz, it is most sensitive to frequencies between 600 and 900 Hz [15–17]. Furthermore, a noise level below 90 decibels (dB) is considered quite safe, and above 100 dB unsafe. All in all, with respect to mining equipment it is to be noted that above-normal noise may make verbal communication, say, between operators and maintenance personnel, impossible. In turn, this can lead to various serious problems including accidents.

5.3.1 Human-Factor Considerations in Equipment Design

In order to have a piece of effective human-compatible equipment, it is essential to consider the relevant human factors during the design stage. At this stage the main objective should be to design a piece of equipment so that it does not subject humans to extreme mental or physical stress or hazards and allows them to perform in the most effective manner.

During the four stages of equipment design shown in Fig. 5.2, design engineers and others should consider human factors from various different perspectives [16, 19]. At the pre-conceptual stage, design-related professionals should systematically define items such as the mission and operational requirements, the functions required to carry out each mission event, and the performance requirements for each mission.

During the concept stage, in addition to the pre-conceptual-stage tasks, the design professionals should also include items such as a preliminary definition of manning and training needs; preliminary task descriptions of users, operators, and maintainers; and analysis for defining the most suitable design method to accomplish each hardware functional assignment. At the pre-design stage, design engineers and associated professionals, in addition to reviewing the analyses of the concept stage, should also conduct time line and link analyses, perform man-machine mock-up studies, etc.

Finally, during the detailed design stage, design-related professionals should consider developing an equipment/system statement, identifying critical skill requirement specifications, creating and evaluating critical man-machine mock-ups, performing link analysis for all important man-machine interfaces, and so on.

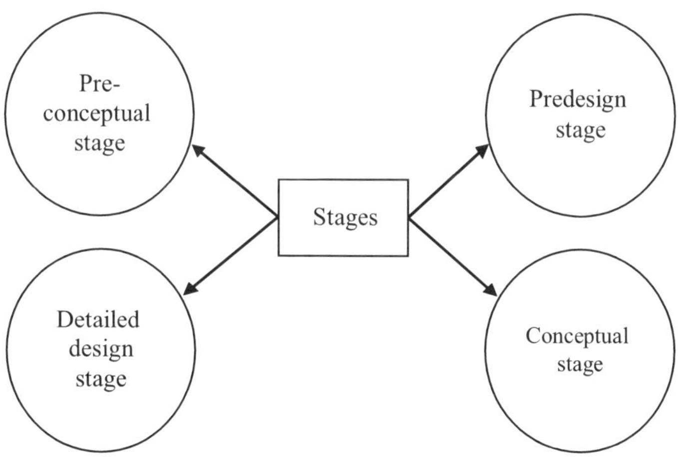

Fig. 5.2 Equipment design stages

5.4 Human-Factor Formulas

There are many mathematical formulas for estimating human-factor-related information. This section presents some of these formulas considered useful for application in the mining industry [16, 17].

5.4.1 Formula I

This formula is concerned with calculating the length of rest periods for humans performing various types of tasks. In mining equipment design, the rest period requirement for humans must be taken into consideration for its ultimate effectiveness.

The duration of a rest period, depending on the task average energy cost, can be estimated by using the following formula [18]:

$$T_r = T_w\,(AE - SEE)\,/\,(AE - R)\ , \tag{5.1}$$

where

- T_r is the required rest period, expressed in minutes;
- T_w is the working time, expressed in minutes;
- SEE is the standard energy expenditure expressed in kilocalories per minute. In the event of having no data, the value of SEE may be taken as 5 Kcal/minute;
- AE is the average energy expenditure expressed in kilocalories per minute of work;
- R is the resting level. Its approximate value is taken as 1.5 Kcal/min.

Example 5.1

A person is performing a maintenance task involving mining equipment for 100 min and his average energy expenditure is 4 Kcal per minute. Calculate the length of the required rest period if the value of $SEE = 3$ Kcal/min.

Substituting the given data values into Eq. (5.1) we get

$$T_r = 100\,(4-3)\,/\,(4-1.5) = 40\,\text{min}\ .$$

Thus, the length of the required rest period is 40 min.

5.4.2 Formula II

As the ability of humans to make visual discrimination depends on factors such as size, exposure time, and illumination, this formula is concerned with estimating character height by considering factors such as illumination, viewing distance, viewing conditions, and the importance of reading accuracy. Thus, the character height is expressed by [21]

$$C_h = \theta D_v + CF_i + CF_v \,, \tag{5.2}$$

where

- C_h is the character height expressed in inches;
- D_v is the viewing distance expressed in inches;
- θ is a constant whose value is taken as 0.0022;
- CF_i is the correction factor for importance or criticality. For critical and noncritical markings, its recommended values are 0.075 and 0, respectively;
- CF_v is the correction factor for viewing and illumination conditions. For its recommended values, see Ref. [22].

5.4.3 Formula III

This formula is concerned with estimating the level of noise/sound intensity in terms of decibels. Thus, the sound-pressure level (SPL), in decibels (dB), is defined by [23, 24]

$$SPL(\text{dB}) = 10\log_{10}\left[\frac{P^2}{P_0^2}\right] , \tag{5.3}$$

where

- P_0^2 is the standard reference sound pressure squared denoting zero decibels. In particular, P_0 is ordinarily the faintest 1,000-Hz tone that an average young individual can hear;
- P is the sound pressure squared of the sound to be measured.

5.4.4 Formula IV

This formula is concerned with estimating the maximum lifting load for a person. The maximum lifting load for a person is expressed by [25]

$$L_m = \alpha\,(IBMS) \,, \tag{5.4}$$

where

L_m is the maximum lifting load for a person,
α is a constant whose values for males and females are 1.1 and 0.95, respectively,
$IBMS$ is the isometric back muscle strength.

5.5 Useful General Human Factors Guidelines for Application in Mining Equipment Design

There are many different aspects of human factors that must be considered during equipment design. Some of the useful human-factor guidelines for application in mining equipment design are as follows [19]:

- Review equipment/system objective with respect to human factors.
- Acquire all appropriate human-factor-design guide and reference documents.
- Develop a human-factor-design checklist for application during design and production phases.
- Use the services of human-factor specialists as the need arises.
- Ensure that the above checklist is used effectively throughout the design and production phases.
- Use appropriate mock-ups for "testing" the effectiveness of user-hardware interface designs.
- Review final equipment production drawings with care with regard to human factors.
- Fabricate a hardware prototype (if possible) for evaluating it under real-life use environments.
- Perform appropriate experiments when cited reference guides fail to provide satisfactory information for making design-related decisions.
- Conduct appropriate field tests of the equipment/system design prior to its approval for final delivery to customers.

5.6 Classifications and Causes of Human Errors Leading to Fatal Accidents in Mines

A study of 794 human errors that resulted in fatal accidents in mines classified these errors under six distinct classifications, as shown in Fig. 5.3 [26].

Causes for the failure to perceive a warning were inadequate inspection technique, neglecting to inspect, obstruction to line of sight, inattention or distraction, masking noise, etc. Three identifiable main causes for the failure to recognize a perceived warning were inadequate information, lack of training, and lack of experience. The main cause for the failure to respond to a recognized warning was

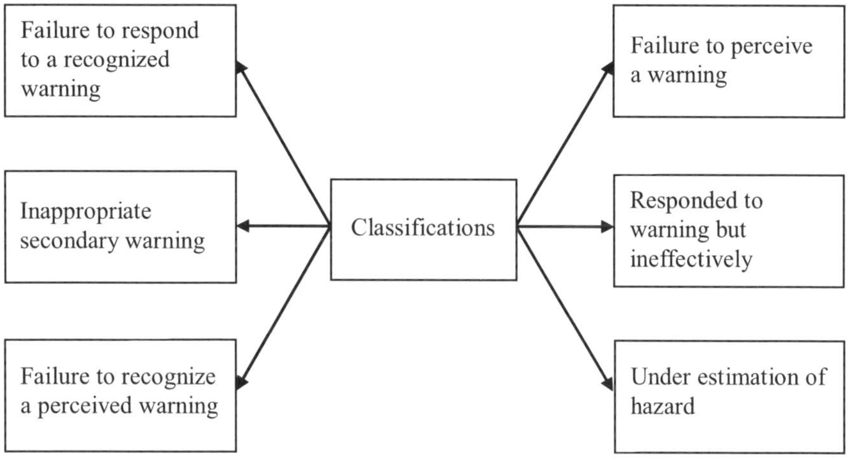

Fig. 5.3 Classifications of human errors leading to fatal accidents in mines

the underestimation of hazard. Three identifiable causes for ineffective response to a warning were negligence or carelessness, inappropriate standard practice, and well-intended but ineffective direct action. Finally, the causes for underestimating a hazard and an inappropriate secondary warning were unidentifiable [26].

5.7 Typical Mining Equipment Maintenance Errors, Factors Contributing to Maintenance Error, and Useful Engineering Design Improvements to Reduce Mining Equipment Maintenance Errors

Some of the typical mining equipment maintenance errors are as follows [27]:

- Failure to follow prescribed instructions and procedures
- Failure to seal or close properly
- Reassemble error
- Failure to lubricate
- Parts installed backward
- Failure to detect while inspecting
- Failure to align, check, or calibrate
- Installation of wrong component
- Failure to act on indicators of problems due to factors such as priorities, workload, or time constraints
- Use of wrong lubricants, greases, or fluids
- Error resulting from failure to complete task due to shift change
- Omitting a part or component

There are many factors that contribute to maintenance-related human error. Some of these factors are as follows [27]:

- Poor manuals
- Poor layout of parts in a compartment
- Confined workspaces
- Inadequate provision for hose and cable management
- Poor task inspection and check-out time
- Inability to make visual inspections
- Lack of proper tools and troubleshooting guides
- Inaccessible components on parts
- Excessive weight of parts or components being manually handled
- Inappropriate placement of parts or components on equipment

Some of the useful engineering design improvement guidelines to reduce mining equipment maintenance errors are as follows [27]:

- Improve fault isolation design by providing built-in test capability, designating test points and procedures, and clearly indicating the direction of fault.
- Use appropriate operational interlocks so that subsystems cannot be turned on if they are wrongly assembled or installed.
- Improve component-equipment interface by designing interfaces so that the component or part can only be installed correctly and providing appropriate mounting pins and other devices to support a part or component while it is being bolted or unbolted.
- Design to facilitate detection of errors.
- Use appropriate decision guides to minimize human guesswork by providing arrows to indicate direction of flow, correct hydraulic pressures, and correct type of lubricants or fluids.
- Improve warning devices, indicators, and readouts to minimize human decision making.

5.8 Types of Chemicals Released in Human-Error-Related Events in the Mining and Manufacturing Industries and Factors Responsible for Failing to Reduce the Occurrence of Human Error in Mines

Past experience shows that human error has played an important role in several large-scale hazardous material events. A study of 3,282 human-error-related events that occurred in the mining and manufacturing industries during the period 1996–2003 revealed that various types of chemicals were released. Most of these chemicals are listed below [28].

- Ammonia
- Chlorine

- Pesticides
- Hydrocarbons
- Acids
- Polymers
- Oxy-organics
- Hetero-organics
- Paints and dyes
- Bases
- Volatile organic compounds
- Polychlorinated biphenyls

Although the number of chemicals released per human-error-released event ranged from 1 to 14, in most of the 3,282 events only one chemical was released.

Some of the important direct or indirect factors in the failure to reduce the occurrence of human errors in mines are as follows [29]:

- Miners are performing their tasks under more difficult environmental, physical, and geo-mining conditions than ever before.
- The greater degree of automation and mechanization in today's mines requires greater understanding, capability, and efficiency from mine workers.
- There is greater worry and mental tension among mine workers due to a greater desire to have more than others.
- Stress in the home is on the rise worldwide.

5.9 Human-Error-Analysis Methods for Application in the Area of Mining

Over the years many methods and techniques have been developed to perform human-error analysis of engineering systems [3]. This section presents three of these methods considered useful for application in the area of mining.

5.9.1 Probability Tree Method

This method is concerned with performing task analysis by diagrammatically representing critical human actions and other events associated with the system under consideration. The branches of the probability tree denote diagrammatic task analysis. More specifically, the outcomes of each event are denoted by the branching limbs of the tree and each of these branching limbs is assigned the probability of occurrence.

The probability tree method has many advantages. It is [30]

- A useful visibility tool;

- A useful tool to decrease the probability of error due to computation resulting from computational simplification;
- A useful tool to readily estimate conditional probability, which may otherwise be obtained through complex probability equations;
- A useful tool to incorporate, with some modifications, factors such as emotional stress, interaction stress, and interaction effects.

The method is demonstrated through the following example.

Example 5.2

A mine worker performs two consecutive tasks, say *m* and *n*, and each of which can either be performed correctly or incorrectly. Furthermore, both tasks are independent of each other (*i.e.*, the performance of task *m* does not affect the performance of task *n* or vice versa), and task *m* is carried out before task *n*.

Obtain an expression for the probability, by developing a probability tree, that the mine worker will not successfully complete the overall mission.

A probability tree for this example is shown in Fig. 5.4. The tree shows that the mine worker first performs task *m* correctly or incorrectly and then proceeds to task *n*, which can also be performed correctly or incorrectly.

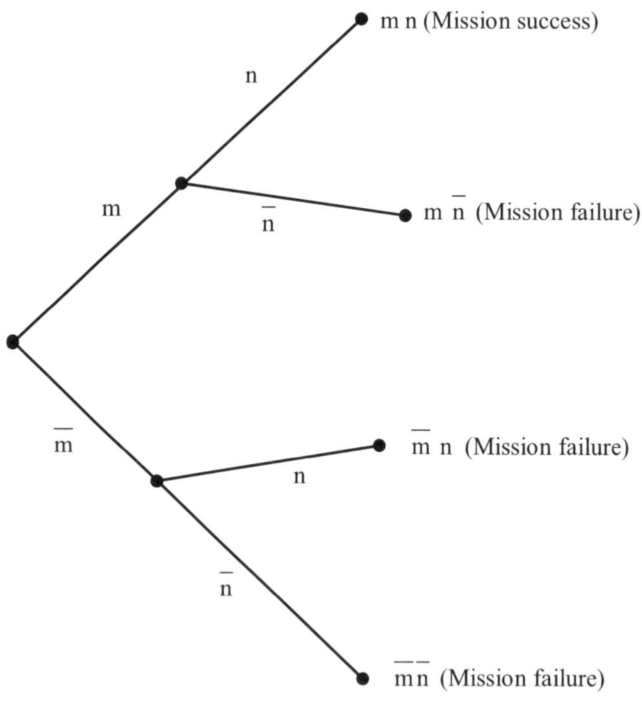

Fig. 5.4 Probability tree for Example 5.2

In Fig. 5.4, m and n with bars denote unsuccessful events and without bars successful events. Other symbols used to obtain the solution expression for the example are defined below.

P_m is the probability of performing task m correctly,
P_n is the probability of performing task n correctly,
$P_{\bar{m}}$ is the probability of performing task m incorrectly,
$P_{\bar{n}}$ is the probability of performing task n incorrectly,
P_{nsu} is the probability of not completing the overall mission.

Using Fig. 5.4, the probability of the mine worker not completing the overall mission is given by

$$P_{nsu} = P_m P_{\bar{n}} + P_{\bar{m}} P_n + P_{\bar{m}} P_{\bar{n}} \,. \tag{5.5}$$

Thus, the expression for the probability that the mine worker will not complete the overall mission is given by Eq. (5.5).

5.9.2 Throughput Ratio Method

This method was developed by the US Navy Electronic Laboratory Center [31]. The ratio generated by the method determines the operability of man-machine interfaces. The term "throughput" implies transmission, because the ratio is expressed in terms of responses per unit time emitted by the equipment/system operator. The throughput ratio in percentage is expressed by

$$R_{\mathrm{mo}} = \left(\frac{\lambda}{\theta} - CF\right)(100) \,, \tag{5.6}$$

where

R_{mo} is the man-machine operability,
CF is the correction factor (*i.e.*, correction for error or out-of-tolerance output),
λ is the number of throughput items generated per unit time,
θ is the number of throughput items to be generated per unit time to meet design expectations.

The correction factor, CF, is expressed by

$$CF = \left[\left(\frac{n_1}{n_2}\right)\left(\frac{\lambda}{\theta}\right)\right]\left[\left(\frac{n_1}{n_2}\right)\left(\frac{\lambda}{\theta}\right) P_{\mathrm{en}}^2 P_{\mathrm{ff}}\right] \,, \tag{5.7}$$

where

n_1 is the number of trials in which the control-display operation is performed incorrectly,
n_2 is the number of trials in which the control-display operation is performed,
P_{en} is the probability that the error will not be detected by the equipment/system operator,
P_{ff} is the probability of function failure due to human error.

All in all, the throughput ratio may be used for various purposes including to demonstrate system acceptability, to establish system feasibility, and to make comparisons of alternative design operabilities [3, 31].

Example 5.3

For the following given values of λ, θ, n_1, n_2, P_{en}, and P_{ff}, calculate the value of the throughput ratio:

$$\lambda = 7\,,\ \theta = 13\,,\ n_1 = 4\,,\ n_2 = 16\,,\ P_{en} = 0.4\,,\ \text{and}\ P_{ff} = 0.8\,.$$

$$CF = \left[\left(\frac{4}{16}\right)\left(\frac{7}{13}\right)\right]\left[\left(\frac{4}{16}\right)\left(\frac{7}{13}\right)(0.4)^2\,(0.8)\right] = 0.00232\,.$$

Substituting the above-calculated value and the given data values into Eq. (5.6) we get

$$R_{mo} = \left(\frac{7}{13} - 0.00232\right)(100) = 53.61\%\,.$$

Thus, the value of the throughput ratio (*i.e.*, the man-machine operability) is 53.61%.

5.9.3 Fault Tree Analysis

This method is widely used to perform various types of reliability analysis and can also be used to perform human error analysis in the mining industry. The method is described in Chap. 3. The following examples demonstrate the application of the method to perform human error analysis in the area of mining.

Example 5.4

Assume that a mine worker is required to perform a certain job, say X, composed of three independent tasks I, J, and K. All three of these tasks must be performed correctly for the successful completion of the job.

Task I is composed of two subtasks I_1 and I_2. If any one of these two subtasks is performed correctly, task I can be accomplished successfully. Similarly, task J is also composed of two subtasks J_1 and J_2. However, both these subtasks must be carried out correctly for the successful completion of task J.

Both subtasks, I_2 and J_2, are made up of three steps each, *i.e.*, i_1, i_2, i_3 and j_1, j_2, j_3, respectively. All steps for each of these two subtasks must be accomplished correctly for subtask success. Develop a fault tree for the event that job X will be performed incorrectly by the mine worker.

Figure 5.5 shows the fault tree for Example 5.4.

Example 5.5

Assume that the probability of occurrence of the basic events (*i.e.*, events denoted by circles) in Fig. 5.5 is 0.02. Calculate the probability of occurrence of the top fault tree event (*i.e.*, job X will be performed incorrectly by the mine worker) by assuming that all the fault tree events occur independently.

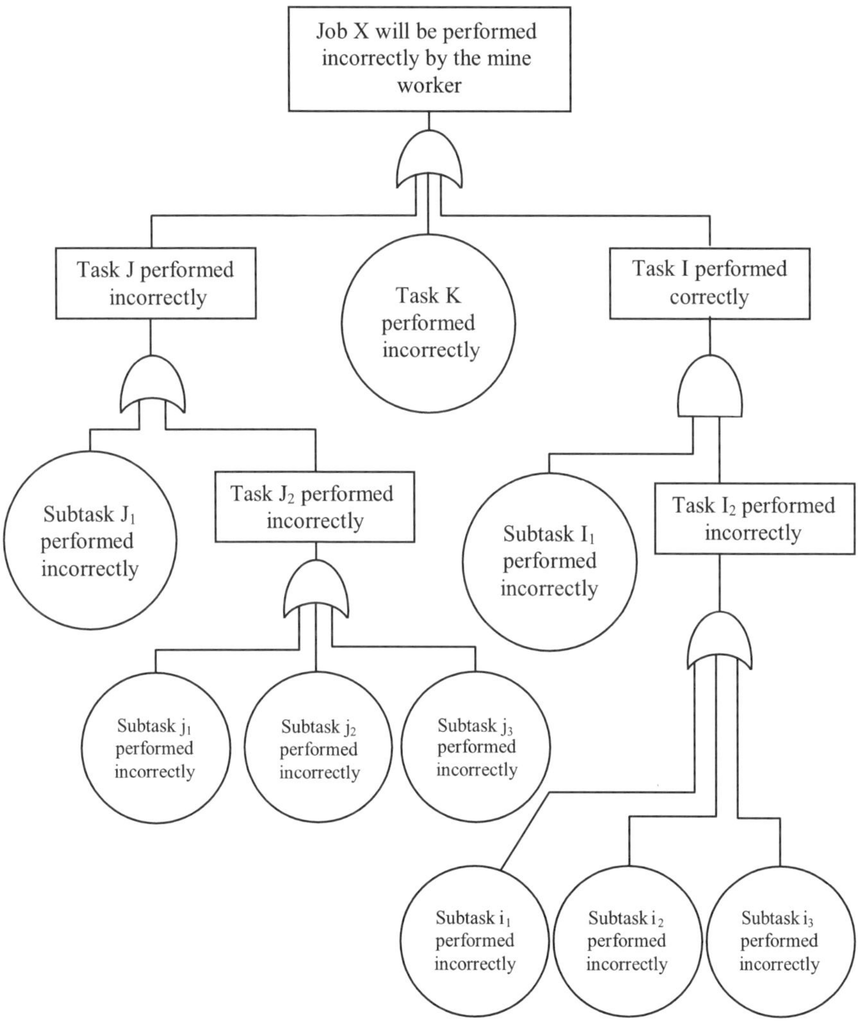

Fig. 5.5 Fault tree for Example 5.4

Using Chap. 3, we perform calculations as follows:
The probability of performing subtask I_2 incorrectly is given by

$$\begin{aligned} P(I_2) &= 1-(1-P(i_1))(1-P(i_2))(1-P(i_3)) \\ &= 1-(1-0.02)(1-0.02)(1-0.02) \\ &= 0.0588 , \end{aligned}$$

where

$P(i_n)$ is the probability of performing step i_n incorrectly, for $n = 1, 2, 3$.

Similarly, the probability of performing subtask J_2 incorrectly is expressed by

$$\begin{aligned} P(J_2) &= 1-(1-P(j_1))(1-P(j_2))(1-P(j_3)) \\ &= 1-(1-0.02)(1-0.02)(1-0.02) \\ &= 0.0588 , \end{aligned}$$

where

$P(j_n)$ is the probability of performing step j_n incorrectly, for $n = 1, 2, 3$.

The probability of performing task J incorrectly is given by

$$\begin{aligned} P(J) &= 1-(1-P(J_1))(1-P(J_2)) \\ P(J) &= 1-(1-0.02)(1-0.0588) \\ &= 0.0776 , \end{aligned}$$

where

$P(J_n)$ is the probability of performing step J_n incorrectly, for $n = 1, 2$.

Using one of the above-calculated values, the probability of performing task I incorrectly is expressed by

$$P(I) = P(I_1)P(I_2) = (0.02)(0.0588) = 0.0012$$

where

$P(I_1)$ is the probability of performing subtask I_1 incorrectly.

Finally, the probability of the mine worker performing job X incorrectly is

$$\begin{aligned} P(X) &= 1-(1-P(J))(1-P(K))(1-P(I)) \\ &= 1-(1-0.0776)(1-0.02)(1-0.0012) \\ &= 0.0971 . \end{aligned}$$

Thus, the probability that job X will be performed incorrectly by the mine worker is 0.0971.

5.10 Problems

1. Discuss the need for human-factor applications in the mining industry.
2. What are the common roadblocks to the introduction of human factors in an organization?
3. Discuss the following human sensory capacities:

   - Sight
   - Touch
   - Noise

4. Discuss human-factor considerations in equipment design.
5. Assume that a mine worker is performing a maintenance task for 100 min and his average energy expenditure is 5 Kcal per minute. Calculate the length of the required rest period if the value of standard energy expenditure (SEE) is 2 Kcal/min.
6. List at least eight general human-factor guidelines considered useful for application in mining equipment design.
7. What are the main causes of human errors leading to fatal accidents in mines?
8. What are the types of chemicals released in human-error-related events in the mining and manufacturing industries.
9. Discuss two human-error-analysis methods considered useful for application in the area of mining.
10. Discuss important factors responsible for failing to reduce the occurrence of human error in mines.

References

1. MIL-STD-721B, Definitions of Effectiveness Terms for Reliability, Maintainability, Human Factors, and Safety, August 1966. Available from the Naval Publications and Forms Center, 5801 Tabor Ave., Philadelphia, PA, USA
2. Meister, D.: Human factors in reliability. In: Ireson, W.G. (ed.): Reliability Handbook. McGraw-Hill, New York, pp. 12.2–12.37 (1966)
3. Dhillon, B.S.: Human Reliability: With Human Factors. Pergamon, New York (1986)
4. Chapanis, A.: Man-Machine Engineering. Wadsworth, Belmont, CA (1965)
5. Williams, H.L.: Reliability evaluation of the human component in human-machine systems. Electric. Manufactur. pp. 78–82 (April 1958)
6. Shapero, A., Cooper, J.I., Rappaport, M., Shaeffer, K.H., Bates, C.J.: Human Engineering Testing and Malfunction Data Collection in Weapon System Programs, WADD Technical Report No. 60–36, Wright-Patterson Air Force Base, Dayton, OH (February 1960)
7. Salvendy, G. (ed.): Handbook of Human Factors and Ergonomics. Wiley, New York (1997)
8. Dhillon, B.S.: Human Reliability and Error in Medical System. World Scientific, River Edge, NJ (2003)
9. Hitchcock, L., Sanders, M. (eds.): Survey of Human Factors in Underground Bituminous Coal Mining Naval Ammunition Depot, Crane, IN (1971)
10. Sanders, M.S., Pay, J.M.: Human Factors in Mining, Information Circular No. 9182, Bureau of Mines, US Department of the Interior, Washington, DC (1991)

11. Conway, E.J., Sanders, M.S.: Recommendations for Human Factors Research and Development Projects in Surface Mining, Report No. OFR 211-83, Bureau of Mines, US Department of the Interior, Washington, DC (1982)
12. Gross, B., Schurick, J.: Hazard Analysis and Safety Economics in Mineral Processing Plants, Report, Canyon Research Group, Westlake Village, CA (1982)
13. Dhillon, B.S.: Bibliography of literature on mining equipment reliability. Microelectron. Reliabil. 26, 1131–1138 (1986)
14. AMCP 706-133, Engineering Design Handbook: Maintainability Engineering Theory and Practice, Prepared by the US Army Material Command, 5001 Eisenhower Ave., Alexandria, VA (1976)
15. AMCP 706-134, Engineering Design Handbook: Maintainability Guide for Design, Prepared by the US Army Material Command, 5001 Eisenhower Ave., Alexandria, VA (1972)
16. Dhillon, B.S.: Advanced Design Concepts for Engineers, Technomic, Lancaster, PA (1998)
17. Dhillon, B.S.: Engineering Design: A Modern Approach, Irwin, Chicago (1996)
18. Altman, J.W., *et al.*: Guidelines to Design of Mechanical Equipment for Maintainability, Report No. ADSD-TR-61-381, United States Air Force (USAF), Aeronautical Systems Division, Dayton, OH (August 1961)
19. Woodson, W.E., Tillman, P., Tillman, B.: Human Factors Design Handbook. McGraw-Hill, New York (1981)
20. Murrell, K.F.H.: Human Performance in Industry. Reinhold, New York (1965)
21. Peters, G.A., Adams, B.B.: Three criteria for readable panel markings. Prod. Eng. 30, 21, 55–57 (1959)
22. Huchingson, R.D.: New Horizons for Human Factors in Design. McGraw-Hill, New York (1981)
23. McCormick, E.J., Sanders, M.S.: Human Factors in Engineering and Design. McGraw-Hill, New York (1982)
24. Adams, J.A.: Human Factors Engineering. MacMillan, New York (1989)
25. Poulsen, E., Jorgensen, K.: Back muscle strength, lifting, and stooped working postures. Appl. Ergonom. 2, 133–137 (1971)
26. Lawrence, A.C.: Human error as a cause of accidents in gold mining. J. Safety Res. 6, 2, 78–88 (1974)
27. Under, R.L., Conway, K.: Impact of maintainability design on injury rates and maintenance costs for underground mining equipment. In: Improving Safety at Small Underground Mines, compiled by R.H. Peters, Special Publication No. 18-94, Bureau of Mines, US Department of the Interior, Washington, DC (1994)
28. Ruckart, P.Z., Burgess, P.A.: Human error and time of occurrence in hazardous material events in mining and manufacturing. J. Hazard. Mater. 142, 747–753 (2007)
29. Mohan, S., Duarte, D.: Cognitive modeling of underground miners response to accidents. In: Proceedings of the Annual Reliability and Maintainability Symposium, 51–55 (2006)
30. Rook, L.W.: Reduction of Human Error in Industrial Production, Report No. SCTM 93-62(14), Sandia Laboratories, Albuquerque, NM (June 1962)
31. Meister, D.: Comparative Analysis of Human Reliability Models, Report No. AD 734–432 (1971). Available from the National Technical Information Service, Springfield, VA 22151

Chapter 6
Mining Equipment Maintainability

6.1 Introduction

In the 1950s underground mining equipment (*e.g.*, in coal mines) basically consisted of various types of simple and rugged machines powered by hydraulics and electric motors. These machines played an instrumental role in cutting, digging, loading, and transporting, say coal, from the mine face to the surface level. The equipment was maintained by maintenance personnel with just basic knowledge of machine design, hydraulics, and electricity and were expected to repair such equipment at the mine site using only simple hand tools.

Over the years, due to the boost in productivity and other factors, the old mining equipment has been transformed into complex and powerful systems. In turn, this has increased the need for better equipment maintainability and knowledge and skills of maintenance workers. Some of the factors directly or indirectly considered in regard to mining equipment maintainability are reducing mean time to repair, lowering life cycle maintenance costs, improving safety, reducing or eliminating altogether the need for maintenance, and reducing the amount, frequency, and complexity of required maintenance tasks.

This chapter presents various important aspects of mining equipment maintainability.

6.2 The Meanings of Mining Equipment Maintainability and Design-induced Maintainability Problems of Mining Equipment

Although the meanings of maintainability may be interpreted in various ways, with regard to mining equipment, maintainability basically means the followings [1]:

- Maintenance considered at design stage
- Effective accessibility of all systems and components

- Predictability of all possible potential failures
- Proper initial startup and commissioning
- Continuous improvement of total machine/equipment
- Serviceability and repairability of all involved systems and components

A study of underground mining equipment with respect to maintainability revealed many design limitations that directly impacted maintenance time, cost, and personnel safety [2, 3]. These problems or limitations are shown in Fig. 6.1. "Accessibility limitations" refers to the inability of maintenance workers in accessing failed or suspected parts for inspection or removal and replacing them. Some of the causes for accessibility problems in mining equipment are as follows [3]:

- Poor access opening size
- Poor layout of parts in a compartment
- Inability to use required tools
- Partially or completely disassembly of equipment to locate fasteners and mechanical interfaces

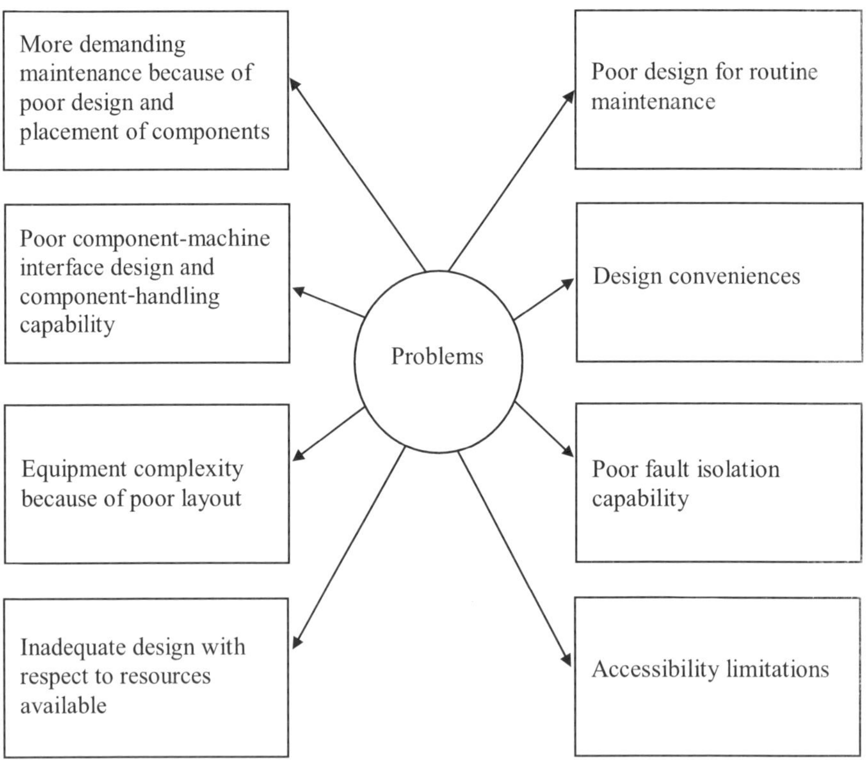

Fig. 6.1 Design-induced maintainability problems of mining equipment

"Poor component-machine interface design and component-handling capability" are basically concerned with the ineffectiveness of the component-handling capability and component-machine interface design. "More demanding maintenance because of poor design and placement of components" is concerned with increased maintenance burden due to poor design and placement of parts or components, subjecting them to impact damage.

"Poor fault isolation capability" is concerned with items such as those listed below:

- Accessing parts or components to carry out visual inspection and to perform appropriate checks
- Difficulty in determining the exact cause and location of a fault or failure
- Lack of appropriate fault indexes
- Limited or existence of absolutely no designed-in fault diagnostic capabilities

"Equipment complexity because of poor layout" is concerned with crowding of parts/components into compartments without paying any attention to factors such as the need to maintain or replace individual items and overlaying hoses and power cables. "Design conveniences" are basically concerned with multiplying the number of connectors, valves, and other high-frequency replacement parts as a design convenience.

"Poor design for routine maintenance" is concerned with tasks such as quickly removing and replacing leaking water lines and hydraulic hoses, performing routine lubrication, removing and replacing failed hydraulic valves, and performing physical and visual inspections. "Inadequate design with respect to resources available" is basically concerned with maintenance workers to "jerry-rig" tools, to handle 45-kg to 450-kg parts, and to use brute human strength for overcoming ineffective component interface design or lack of requisite tools.

6.3 Advantages of Improved Mining Equipment Maintainability Design

The main objectives of the maintainability engineering application to mining equipment are to increase efficiency and safety and to reduce maintenance cost. This starts with improving the mining equipment design. Nonetheless, although the application of maintainability engineering will not eliminate the need for service and maintenance of mining equipment, it will certainly provide advantages such as the following [2]:

- Reduction in maintenance-related injuries
- Reduction in time required to accomplish scheduled and unscheduled maintenance
- Reduction in maintenance-related human errors
- Reduction in incorrect installations

- Improvements in troubleshooting performance
- Minimization of the unscheduled maintenance frequency due to improvements in accessibility for inspection and servicing
- Reduction in the need for the training of maintenance personnel
- Improvement in postmaintenance inspection

6.4 Mining Equipment Maintainability Design Characteristics

Many maintainability design characteristics must be considered with care during the mining equipment design phase. These characteristics include those factors or features that play an important role in reducing the mining equipment downtime and unavailability. These features or factors include standardization, interchangeability, accessibility, and safety. Each of these features is described below, separately [4–8].

6.4.1 Standardization

This is an important design feature and is basically concerned with restricting to a minimum the variety of parts and components that can be used to meet the equipment/product requirements. Some of the primary goals of standardization are to maximize the use of interchangeable and standard parts, minimize the use of different types of parts, maximize the use of common parts in different products, and minimize the number of different models and makes of equipment in use.

Some of the main advantages of standardization are improvement in equipment reliability and maintainability, reduction in incorrect use of parts, reduction in manufacturing costs, design time, and maintenance time and cost, and reduction in the probability of occurrence of accidents stemming from wrong or unclear procedures [7, 8].

6.4.2 Interchangeability

Interchangeability is an important maintainability design factor, and it simply means that a given item can be replaced by any similar item, and the replacing item can perform the specified functions of the replaced item in an effective manner. There are two types of interchangeability: physical interchangeability and functional interchangeability. In the case of physical interchangeability, two items can be connected, mounted, and used in the same location and in the same way. Similarly, in the case of functional interchangeability, two given items serve the same function.

Needless to say, maximum interchangeability can only be achieved if the design professionals carefully consider items such as providing effective information in the task instructions and physical similarities (*e.g.*, shape and size).

6.4.3 Accessibility

Accessibility may simply be described as the relative ease with which an item can be reached for repair, replacement, or service. As per past experiences, lack of accessibility is an important maintainability problem and often a cause of poor maintenance [4]. Some of the important factors that affect maintainability are as follows [4, 5, 8]:

- Environment and location of the item to be accessed
- Frequency with which the access opening is entered
- The type of maintenance tasks to be performed through the access opening
- Work clearances necessary for carrying out the required tasks
- Types of accessories and tools required to carry out the tasks
- Distance to be reached to access the item in question
- Visual requirements of personnel performing the tasks
- The degree of danger involved in using the access opening
- Specified time requirements for carrying out the tasks

6.4.4 Safety

This is an important maintainability design factor because maintenance personnel performing various types of tasks may be exposed to hazardous conditions. These conditions could be the result of poor consideration given to the safety aspect during the design phase. Nonetheless, some of the human safety guidelines are to install fail-safe devices as considered appropriate, fit all access openings with appropriate fillets, study the potential sources of injury by electric shock, install items requiring maintenance in such a way that hazard in accessing them is minimized, and provide appropriate emergency doors.

6.5 Maintainability Measures for Mining Equipment

Normally quantitative maintainability specifications are based on desired limiting conditions imposed on equipment maintenance labor-hours, downtime, and so on. Thus, in order to have an effective and mining equipment maintainability, it is important to have knowledge of various maintainability parameters or measures during the design phase. These measures include mean time to repair, mean preventive maintenance time, and the probability of completing repair in a given time interval (*i.e.*, the maintainability function). All of these measures are presented below, separately [4, 7–10].

6.5.1 Mean Time to Repair

This is probably the most widely used measure or parameter in maintainability analysis and is also referred to as mean corrective maintenance time. Normally, probability distributions such as normal, log-normal, and exponential are used to represent corrective maintenance times.

The system/equipment mean time to repair (MTTR) is defined by

$$MTTR = \left[\sum_{i=1}^{m} \lambda_i T_i\right] / \sum_{i=1}^{m} \lambda_i \,, \tag{6.1}$$

where

- m is the number of units;
- λ_i is the constant failure rate of unit i; for $i = 1, 2, 3, 4, \ldots, m$;
- T_i is the corrective maintenance or repair time required to repair unit i; for $i = 1, 2, 3, 4, \ldots, m$.

Example 6.1

Assume that a piece of mining equipment is composed of four replaceable subsystems 1, 2, 3, and 4 with constant failure rates $\lambda_1 = 0.0004$ failures/h, $\lambda_2 = 0.0005$ failures/h, $\lambda_3 = 0.0007$ failures/h, and $\lambda_4 = 0.0008$ failures/h, respectively. Corrective maintenance times associated with subsystems 1, 2, 3, and 4 are $T_1 = 2$ h, $T_2 = 3$ h, $T_3 = 1.5$ h, and $T_4 = 0.5$ h, respectively.

Calculate the mining equipment mean time to repair.

Inserting the specified data values into Eq. (6.1) we get

$$\begin{aligned} MTTR &= \frac{(0.0004)(2) + (0.0005)(3) + (0.0007)(1.5) + (0.0008)(0.5)}{(0.0004) + (0.0005) + (0.0007) + (0.0008)} \\ &= 1.565\,\text{h}\,. \end{aligned}$$

Thus, the mining equipment mean time to repair is 1.565 h.

6.5.2 Mean Preventive Maintenance Time

In order to keep mining equipment at a specified performance level, various preventive maintenance-associated activities such as calibrations, tuning, and inspections are performed. A well-planned and well-executed preventive maintenance program can be helpful in reducing, directly or indirectly, mining equipment downtime and improving its performance. Nonetheless, the equipment mean preventive maintenance time (*MPMT*) is defined by

$$MPMT = \left[\sum_{i=1}^{m} ET_i F_i\right] / \sum_{i=1}^{m} F_i \,, \tag{6.2}$$

where

m is the number of preventive maintenance tasks;
ET_i is the elapsed time for preventive maintenance task i; for $i = 1, 23, \ldots, m$;
F_i is the frequency of preventive maintenance task i; for $i = 1, 2, 3, \ldots, m$.

In calculating the value of *MPMT*, it is to be noted that if the frequencies, F_i, are specified in maintenance tasks per hour, then the ET_i must be expressed in hours.

6.5.3 Maintainability Function

The maintainability function is used to compute the probability of accomplishing a repair in a given time interval; it is defined by

$$m(t) = \int_0^t f_r(t)\,\mathrm{d}t \,, \tag{6.3}$$

where

$m(t)$ is the maintainability function,
$f_r(t)$ is the repair time probability density function,
t is the variable repair time.

Maintainability functions for normal and exponential repair time distributions are obtained as follows:

Normal Distribution

The distribution repair time probability density function is expressed by

$$f_r(t) = \frac{1}{\sigma\sqrt{2\pi}} \exp\left[-\frac{1}{2}\left\{\frac{t-\mu}{\sigma}\right\}^2\right] , \tag{6.4}$$

where

μ is the mean of repair times,
σ is the standard deviation of the variable repair time t around the mean μ.

Inserting Eq. (6.4) into Eq. (6.3) we obtain

$$m(t) = \frac{1}{\sigma\sqrt{2\pi}} \int_{-\infty}^{t} \exp\left[-\frac{1}{2}\left\{\frac{t-\mu}{\sigma}\right\}^2\right] \mathrm{d}t \,. \tag{6.5}$$

The mean of repair times is given by

$$\mu = \sum_{i=1}^{n} \frac{t_i}{n} \,, \tag{6.6}$$

where

n is the number of repair times,
t_i is the repair time i, for $i = 1, 2, 3, \ldots, n$.

The standard deviation is expressed by

$$\sigma = \left[\sum_{i=1}^{n} (t_i - \mu)^2 / (n-1)\right]^{1/2} \,. \tag{6.7}$$

Exponential Distribution

The distribution repair time probability density function is defined by

$$f_r(t) = \theta \exp(-\theta t) \,, \tag{6.8}$$

where

θ is the constant repair rate (*i.e.*, the reciprocal of the mean time to repair, *MTTR*),
t is the variable repair time.

Substituting Eq. (6.8) into Eq. (6.3) yields

$$m(t) = \int_{0}^{t} \theta \exp(-\theta t)\,\mathrm{d}t = 1 - \exp(-\theta t) \,. \tag{6.9}$$

Since $\theta = \frac{1}{MTTR}$, Eq. (6.9) becomes

$$m(t) = 1 - \exp\left(-\frac{t}{MTTR}\right) \,. \tag{6.10}$$

Example 6.2

Assume that the repair times of a piece of mining equipment are exponentially distributed with a mean value of 2 h (*i.e.*, $MTTR = 2\,\text{h}$). Calculate the probability of completing a repair action in 4 h.

Substituting the specified data values into Eq. (6.10) we get

$$m(4) = 1 - \exp\left(-\frac{4}{2}\right) = 0.8647 .$$

This means that the probability of completing a repair action within 4 h is 0.8647.

6.6 Common Maintainability Design Errors and Useful Maintainability Design Guidelines for Mining Equipment

During equipment design, often various types of errors are made that adversely affect equipment maintainability. Therefore, it is important that careful attention be given during the design of mining equipment to such errors that may impact its maintainability. Nonetheless, some of the common maintainability-related design errors are as follows [7, 8, 11]:

- Locating adjustable screws close to an exposed power supply terminal or a hot component
- Placing an adjustment out of arm's reach
- Providing unreliable built-in test equipment
- Using access doors with numerous small screws
- Placing low-reliability items beneath other items
- Omitting appropriate handles
- Placing screwdriver-related adjustments underneath modules
- Placing removable parts in such a way that they cannot be dismantled without taking the entire unit from its case
- Providing inadequate space for maintenance workers to get their gloved hands into the unit to carry out necessary adjustments
- Placing adjustable screws in locations difficult for maintenance workers to find

Over the years professionals working in the area of maintainability engineering have developed various guidelines for use during the equipment design phase to improve the effectiveness of equipment maintainability. These guidelines can equally be applied to mining equipment. Some of these guidelines are as follows [11]:

- Design so as to minimize the need for tools and adjustments.
- Design so as to minimize the need for maintenance skills.
- Provide effective troubleshooting methods.
- Design for safety.

- Provide test points at appropriate places.
- Use color coding.
- Provide for effective visual inspection.
- Avoid the use of large cable connectors.
- Make use of captive-type chassis fasteners.
- Label units.
- Use standard interchangeable parts.
- Make use of plug-in instead of solder-in modules.
- Provide appropriate handles on heavy parts for ease of handling.
- Group subsystems.

6.7 Conclusions: State of Maintainability in the Underground Mining Industry

A study conducted by the US Bureau of Mines analyzed various aspects of equipment maintainability in the mining industry [3]. Some of its conclusions are as follows [3]:

- There is little evidence of the systematic application of maintainability principles in the design of operational underground coal mining equipment.
- Similarly, there is little evidence of the systematic application of human-factor principles in the design of this equipment in regard to maintenance.
- Although in the case of some machines heavy maintenance tasks could be carried out on the surface or in high-roof underground shops equipped with the requisite lifting devices, at the mine face it is extremely difficult, time consuming, and risky to perform the same tasks.
- In the case of newer and larger mining machines, there is a significant increase in task complexity and completion times with respect to maintenance.
- With the exception of some machines, task completion times for the ten most frequently performed maintenance tasks could be reduced by approx. 10 to 30% with simple improvements in equipment design.
- Maintenance risk could be reduced quite substantially with the application of accepted human-factor engineering design standards and criteria.

6.8 Problems

1. Discuss the meanings of mining equipment maintainability.
2. List the five most important design-induced maintainability problems of mining equipment.
3. What are the advantages of improved mining equipment maintainability design?
4. Discuss the following items:

- Interchangeability
- Accessibility

5. Define mean preventive maintenance time.
6. Assume that a mining system is made up of three replaceable subsystems 1, 2, and 3 with constant failure rates $\lambda_1 = 0.005$ failures/h, $\lambda_2 = 0.007$ failures/h, and $\lambda_3 = 0.009$ failures/h, respectively. Corrective maintenance times associated with subsystems 1, 2, and 3 are $T_1 = 0.5$ h, $T_2 = 2$ h, and $T_3 = 3$ h, respectively. Calculate the mining system mean time to repair.
7. Define maintainability function.
8. Prove Eq. (6.9).
9. What are the commonly occurring maintainability design errors?
10. List at least ten design guidelines useful for improving mining equipment maintainability.

References

1. Leugner, T.: Developing a total productive maintenance (TPM) programme. Quarry Manage. pp. 21–23 (August 1996)
2. Conway, E.J., Unger, R.L.: Maintainability Design of Underground Mining Equipment, Report No. BuMines OFR 39-91-V1, US Bureau of Mines, Washington, DC (1988)
3. Unger, R.L., Conway, K.: Impact of maintainability design on injury rates and maintenance costs for underground mining equipment. In: Improving Safety at Small Underground Mines, compiled by R.H. Peters, Special Publication No. 18-94, US Bureau of Mines, Washington, DC (1994)
4. AMCP 706 – 133, Engineering Design Handbook: Maintainability Engineering Theory and Practice, prepared by the Army Material Command, Department of the Army, Washington, DC (1976)
5. AMCP 706 – 134, Maintainability Guide for Design, prepared by the Army Material Command, Department of the Army, Washington, DC (October 1972)
6. Wohl, J.G.: Why Design for Maintainability? IRE Trans. Hum. Factors Electron. HFE2, 87–92 (1961)
7. Dhillon, B.S.: Design Reliability: Fundamentals and Applications. CRC, Boca Raton, FL (1999)
8. Dhillon, B.S.: Engineering Maintainability: How to Design for Reliability and Easy Maintenance. Gulf, Houston, TX (1999)
9. Grant-Ireson, W., Coombs, C.F. (eds.): Handbook of Reliability Engineering and Management. McGraw-Hill, New York (1988)
10. Blanchard, B.S., Verma, D., Peterson, E.L.: Maintainability. Wiley, New York (1995)
11. Pecht, M. (ed.): Product Reliability, Maintainability, and Supportability Handbook. CRC. Boca Raton, FL (1995)

Chapter 7
Mining Equipment Reliability and Maintainability Testing

7.1 Introduction

Testing is an important element of any engineering product development program. It may simply be stated as subjecting a product to conditions that highlight its weaknesses, behavior characteristics, and modes of failure.

Reliability testing is an important element of testing and is basically concerned with obtaining information regarding failures, in particular the product/equipment's tendency to fail as well as the failure consequences. A good reliability test program is one that requires a minimal amount of testing and provides the maximum amount of information on failures [1, 2].

The main objective of maintainability testing and demonstration is to verify the maintainability features that have been designed and built into a piece of equipment/product [3]. In addition, maintainability testing and demonstration provides customers with confidence, prior to making production-related commitments, that the piece of equipment/product under consideration meets the specified maintainability-associated requirements.

This chapter presents various important aspects of reliability and maintainability testing considered useful for application in the areas of mining.

7.2 Reliability Test Classifications

Reliability tests may be grouped into three categories, as shown in Fig. 7.1 [4]. These are reliability development and demonstration testing, qualification and acceptance testing, and operational testing.

Three main objectives of reliability development and demonstration testing are as follows:

- To identify necessary changes in design.

B.S. Dhillon, *Mining Equipment Reliability, Maintainability, and Safety*

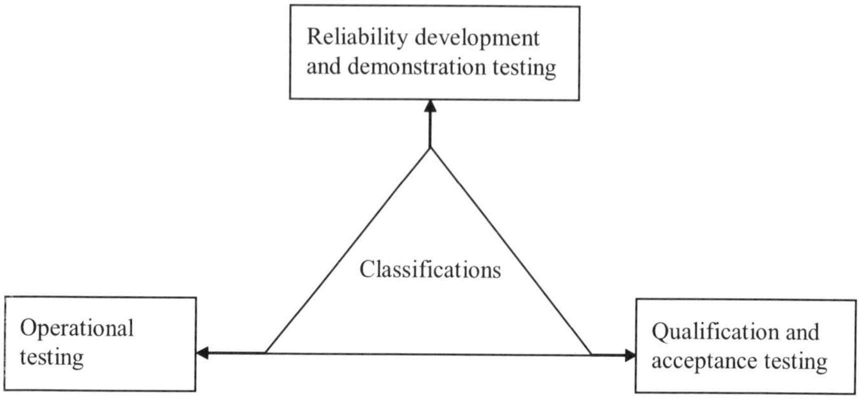

Fig. 7.1 Reliability test classifications

- To determine if there is any need to improve design to meet reliability specifications.
- To verify improvements in design reliability.

Note that the nature of this type of testing depends on various factors including the type of system/equipment/subsystem being investigated and the level of complexity under consideration.

Two main objectives of the qualification and acceptance testing are as follows:

- To determine if the item under consideration should be accepted or rejected individually or on a lot basis.
- To determine if the design under consideration qualifies for its specified objective.

Note that the above two objectives differ from the objectives of other reliability tests, particularly with respect to the accept-or-reject approach.

Finally, three main objectives of the operational testing are as follows:

- To verify the results of reliability analyses performed during the system/equipment design and development phase.
- To provide data for use in subsequent activities.
- To provide data indicating necessary changes to operational policies and procedures with respect to reliability and maintainability.

7.3 Success Testing

This type of testing is normally used in receiving inspection and in engineering test laboratories where a no-failure test is specified; thus it could be quite useful for application in the mining industry. Nonetheless, the main goal of success testing is to ensure that a specified reliability level is achieved at a given confidence level.

Thus, in this case for zero failures, the lower 100 $(1-\alpha)$% confidence limit on the desired reliability level can be expressed as follows [5]:

$$R_L = \alpha^{1/m}\,, \tag{7.1}$$

where

m is the total number of items/units placed on test,
α is the consumer's risk or the level of significance.

Thus, with 100 $(1-\alpha)$% confidence, we may write

$$R_L \le R_t\,, \tag{7.2}$$

where

R_t is the true or actual reliability.

Taking the natural logarithms of both sides of Eq. (7.1) yields

$$\ln R_L = \frac{1}{m}\ln\alpha\,. \tag{7.3}$$

Rearranging Eq. (7.3) we get

$$m = \frac{\ln\alpha}{\ln R_L}\,. \tag{7.4}$$

The desired level of confidence, C, may be expressed as follows:

$$C = 1-\alpha\,. \tag{7.5}$$

Rearranging Eq. (7.5) we obtain

$$\alpha = 1-C\,. \tag{7.6}$$

Substituting Eqs. (7.2) and (7.6) into Eq. (7.4) we get

$$m = \frac{\ln(1-C)}{\ln R_t}\,. \tag{7.7}$$

Thus, Eq. (7.7) can be used to determine the total number of items/units to be tested for a given reliability and confidence level.

Example 7.1

Assume that a manufacturer of a certain part used in a piece of mining equipment is required to demonstrate 95% reliability of that part at a 90% confidence level. Determine the total number of parts to be tested when only zero failures are allowed.

Substituting the given data values into Eq. (7.7) we get

$$m = \frac{\ln(1-0.90)}{\ln 0.95} \cong 45\,\text{parts}\,.$$

This means a total of 45 parts must be tested.

7.4 Accelerated Testing

Accelerated testing is concerned with reducing the duration of the test time by varying parameters such as frequency of cycling, voltage above regular levels, or temperature or by simply conducting sudden-death testing. There are a number of ways to perform accelerated testing including those listed below [6–11].

- Perform sudden-death testing.
- Increase the sample size.
- Increase the test severity.

When performing sudden-death testing, the test sample is divided into various groups containing an equal number of items or units, and all items/units in each of these groups start their operation simultaneously. Whenever any one of the group units fails, all units are considered failed and the testing of nonfailed units is terminated immediately. Furthermore, in the event of failure of the first unit in the last and final group, the entire testing process is terminated.

With respect to increasing the sample size, the size of the sample is increased when the item/unit life distribution does not exhibit a wear-out characteristic during its (*i.e.*, item's) predicted lifespan. Note that the increase in the sample size decreases the test time as the test time is inversely proportional to the sample size.

Finally, increasing the test severity involves increasing the stress acting on the test unit/item. The stress may be classified under two distinct areas: application and operational. The application area includes items such as voltage, self-generated heat, and current; the operational area includes humidity and temperature. The following equation can be used to calculate the acceleration factor for this type of accelerated testing [10–13]:

$$F_a = \exp\left[-(E/k)\left\{\frac{1}{T_{ac}} - \frac{1}{T_{us}}\right\}\right], \tag{7.8}$$

where

F_a is the acceleration factor,
T_{ac} is the acceleration temperature expressed in degrees Kelvin,
T_{us} is the use temperature expressed in degrees Kelvin,
E is the activation energy, and its value is taken as 0.5 eV,
k is the Boltzmann constant, and its value is taken as 0.00008623 eV/degree.

Example 7.2

A sample of 20 identical engineering items used in mines were tested to failure at a temperature of 130 °C, and their mean time between failures (*MTBF*) was 2500 h. Calculate the *MTBF* of these items at the normal operating temperature of 85 °C.

Using the given data values we get

$$T_{ac} = 273 + 130 = 403 \text{ degrees Kelvin}$$

and

$$T_{us} = 273 + 85 = 358 \text{ degrees Kelvin}\,.$$

Substituting the above-calculated values and the other given data values into Eq. (7.8) yields

$$F_a = \exp\left[-\left(\frac{0.5}{0.00008623}\right)\left\{\frac{1}{403} - \frac{1}{358}\right\}\right] = 6.10\,.$$

Thus, the *MTBF* of the engineering items, used in mines, at 85 °C is

$$MTBF = (6.10)(2500) = 15{,}250\,\text{h}\,,$$

where

MTBF is the mean time between failures of the engineering items at 85 °C.

7.5 Confidence Interval Estimates for Mining Equipment Mean Time Between Failures

Usually, in reliability studies conducted for practical purposes, including in the area of mining, times to item failure are assumed to be exponentially distributed. Consequently, the item failure rate becomes constant and, in turn, the item mean time to failure (*MTTF*) or mean time between failures (*MTBF*) is simply the reciprocal of the item failure rate.

Thus, in testing a sample of items with exponentially distributed times to failure, one can make a point estimate of the *MTBF*. Unfortunately, this approach does not provide any surety of measurement because it provides only an incomplete picture. Therefore, it would be more meaningful if we said, for example, that after testing a sample of identical items for T hours, n number of failures occurred and the actual *MTBF* lies somewhere between certain upper and lower limits with a certain degree of confidence.

The confidence interval on *MTBF* can be calculated with the aid of a χ^2 (chi-square) distribution. The following notation is used to obtain chi-square values:

$$\chi^2(p, df)\,, \tag{7.9}$$

where

df is the degrees of freedom,
p is a quantity, which is a function of the confidence coefficient.

The symbols listed below are used in subsequent associated formulas [2, 4]:

α	is the acceptable error risk,
θ	is the mean time between failures (*MTBF*),
n	is the number of items that were placed on test at time $t = 0$,
K	is the number of failures accumulated by time t^*, where t^* is the life test termination time,
K^*	is the number of pre-assigned failures,
$C = 1 - \alpha$	is the confidence level.

Confidence intervals are estimated under the following two conditions:

- Testing is terminated at a pre-assigned number of failures, K^*.
- Testing is terminated at a pre-assigned time, t^*.

The following two formulas are used to calculate upper and lower limits for the above two conditions, respectively [2, 4]:

- Pre-assigned number of failures, K^*:

$$\left[\frac{2Y}{\chi^2\left(\frac{\alpha}{2}, 2K\right)}, \frac{2Y}{\chi^2\left(1-\frac{\alpha}{2}, 2K\right)}\right] ; \tag{7.10}$$

- Pre-assigned truncation time, t^*:

$$\left[\frac{2Y}{\chi^2\left(\frac{\alpha}{2}, 2K+2\right)}, \frac{2Y}{\chi^2\left(1-\frac{\alpha}{2}, 2K\right)}\right] . \tag{7.11}$$

The value of Y is determined by the type of test: replacement test (*i.e.*, the failed item/unit is repaired or replaced) or nonreplacement test. Thus, for the replacement test, the value of Y is expressed by

$$Y = nt^* . \tag{7.12}$$

Similarly, for the nonreplacement test, the value of Y is expressed by

$$Y = (n-K)t^* + \sum_{j=1}^{K} t_j , \tag{7.13}$$

where

t_j is the jth failure time.

In the case of censored items/units (*i.e.*, withdrawal or loss of nonfailed items/units) the value of Y, for replaced failed units/items but nonreplacement of censored

Table 7.1 Chi-square distribution values

Degrees of freedom (df)	Probability					
	0.99	**0.95**	**0.9**	**0.5**	**0.05**	**0.01**
2	0.02	0.1	0.21	1.38	5.99	9.21
4	0.29	0.71	1.06	3.35	9.44	13.27
6	0.87	1.63	2.2	5.34	12.59	16.81
8	1.64	2.73	3.49	7.34	15.50	20.09
10	2.55	3.94	4.86	9.34	18.30	23.20
12	3.57	5.22	6.3	11.34	21.02	26.21
14	4.66	6.57	7.79	13.33	23.68	29.14
16	5.81	7.96	9.31	15.33	26.29	32
18	7.01	9.39	10.86	17.33	28.86	34.80
20	8.26	10.85	12.44	19.33	31.41	37.56
22	9.54	12.33	14.04	21.33	33.92	40.28
24	10.85	13.84	15.65	23.33	36.41	42.98
26	12.19	15.37	17.29	25.33	38.88	45.64
28	13.56	16.92	18.93	27.33	41.33	48.27
30	14.95	18.49	20.59	29.33	43.77	50.89
40	22.14	26.50	29.06	39.33	55.75	63.70
50	29.68	34.76	37.70	49.33	67.50	76.16
60	37.46	43.18	46.47	59.33	79.08	88.39

items/units, is expressed by

$$Y = (n-m)t^* + \sum_{i=1}^{m} t_i \,, \tag{7.14}$$

where

t_i is the ith censorship time,
m is the number of censored items/units.

Similarly, the value of Y, for nonreplaced failed and censored items/units is expressed by

$$Y = (n-m-K)t^* + \sum_{i=1}^{m} t_i + \sum_{j=1}^{K} t_j \,. \tag{7.15}$$

Table 7.1 presents some tabulated values of $\chi^2(p, df)$ [1,2].

Example 7.3

Assume that a total of 20 identical items used in mines were placed on test at time $t = 0$ and none of the failed items were replaced. The test was terminated after 200 h. Five items failed after 50, 70, 120, 130, and 150 h of operation.

Estimate the items' *MTBF* and its upper and lower limits with 90% confidence level.

Substituting the given data values into Eq. (7.13) we get

$$Y = (20-5)200 + (50+70+120+130+150) = 3520\,\mathrm{h}\,.$$

Thus, the items' *MTBF* is expressed by

$$\theta = \frac{3520}{5} = 704\,\mathrm{h}\,.$$

Substituting the specified and calculated values into relationship (7.11) and using Table 7.1, we get the following upper and lower limits for the items' *MTBF*:

$$\begin{aligned}
\mathit{Upper}\ \text{limit} &= \frac{2\,(3520)}{\chi^2\,(0.95,10)}\\
&= \frac{2\,(3520)}{3.94}\\
&= 1786.8\,\mathrm{h}\,;
\end{aligned}$$

$$\begin{aligned}
\mathit{Lower}\ \text{limit} &= \frac{2\,(3520)}{\chi^2\,(0.05,12)}\\
&= \frac{2\,(3520)}{21.02}\\
&= 334.9\,\mathrm{h}\,.
\end{aligned}$$

Thus, the items' *MTBF* is 704 h. Its (*i.e.*, *MTBF*'s) upper and lower limits with a 90% confidence level are 1786.8 h and 334.9 h, respectively. In other words, the true *MTBF* of the items with a 90% confidence level will lie within 334.9 h and 1786.8 h or $334.9 \leq \theta \leq 1786.8$.

7.6 Documents on Reliability Testing

Over the years, many publications have appeared on reliability testing that could be directly or indirectly useful in the reliability testing of mining equipment. Some of the important ones are as follows:

- MIL-STD-781D, Reliability Design, Qualification and Production Acceptance Tests: Exponential Distribution, US Department of Defense, Washington, DC
- IEC 605, Equipment Reliability Testing, International Electrotechnical Commission (IEC), Geneva, Switzerland
- MIL-STD-2074, Failure Classification for Reliability Testing, US Department of Defense, Washington, DC
- IEC 1123, Reliability Testing Compliance Test Plans for Success Ratio, International Electrotechnical Commission (IEC), Geneva, Switzerland

- MIL-HDBK-781, Reliability Test Methods, Plans, and Environments for Engineering Development, Qualification, and Production, US Department of Defense, Washington, DC
- MIL-HDBK-H108, Sampling Procedures and Tables for Life and Reliability Testing (Based on Exponential Distribution), US Department of Defense, Washington, DC
- MIL-STD-202F, Test Methods for Electronic and Electrical Component Parts, US Department of Defense, Washington, DC
- MIL-STD-2165, Testability Program for Electronic Systems and Equipment, US Department of Defense, Washington, DC

7.7 Planning and Control Requirements for Mining Equipment Maintainability Testing and Demonstration

These requirements may be divided into the following six distinct categories [14, 15]:

- **Following MIL-STD-471 guidelines.** These guidelines were developed by the United States Department of Defense in 1966 for manufacturers to carefully consider in the planning and control of equipment maintainability demonstrations. They cover areas such as establishing test teams, selecting a test method, and data collection and selection, performance, and sampling of corrective and preventive maintenance tasks.
- **Providing appropriate manpower.** This is concerned with ensuring that individuals performing maintainability demonstrations have backgrounds and skill levels similar to those of a product's final users, operating, and maintenance personnel.
- **Developing a demonstration plan.** This is concerned with developing a demonstration plan that conforms to the specifics described in MIL-STD-471 [16] and covers areas such as test planning, administration, and control; test conditions; and test analysis, documentation, and reporting.
- **Specifying appropriate parameters.** As the basic purpose of a formal maintainability demonstration is to verify compliance with defined parameters, it is essential to have specifications for demonstration parameters in quantitative terms. A typical example of a measurable time parameter is mean time to repair (*MTTR*).
- **Developing a demonstration model.** This model is used to demonstrate a product's proposed quantitative parameters and qualitative design features for maintainability.
- **Taking environment into consideration.** Environment is an important factor in maintainability testing as equipment downtime may vary significantly between laboratory-controlled conditions and actual operational conditions. Therefore, it is essential to take into consideration factors such as test facilities, support resource needs, and limitation simulations.

7.8 Test Methods to Obtain Maintainability-related Test Data for Mining Equipment

There are numerous points in mining equipment life cycle and in related maintainability program tasks that need test data. This type of data may be necessary to make decisions about maintainability design requirements, to evaluate equipment life cycle maintenance support, or for administrative and logistic control to update corrective actions or modifications.

Test methods that can provide such data may be grouped under six categories as follows [15]:

- **Closed-loop tests.** These tests generate useful information to evaluate equipment-design effectiveness, tolerance adequacy, performance, and other important issues.
- **Functional tests.** These tests simulate normal operating conditions to establish a product's or piece of equipment's state of readiness to perform its stated mission.
- **Static tests.** These tests provide information on the transient behavior of the product under consideration.
- **Dynamic tests.** These tests simulate a typical application of a given piece of equipment so that each and every item involved can be checked properly.
- **Open-loop tests.** These tests represent a refinement of static and dynamic tests and usually provide better maintenance-related information than closed-loop tests because they make a direct observation of the system transfer function without the modifying influence of feedback.
- **Marginal tests.** These tests are used to isolate potential difficulties through the simulation of abnormal operational conditions.

7.9 Test Methods for Demonstrating Diverse Maintainability Parameters

The main objective of demonstrating maintainability is to determine whether a development program or a manufacturer has satisfied qualitative and quantitative maintainability requirements. A successful maintainability demonstration depends on many factors including quality of equipment design for testability, quality of written maintenance manuals, and quality of training of repair technicians.

A total of 11 test methods addressing many diverse maintainability parameters considered useful for application in the mining industry are shown in Fig. 7.2 [16, 17]:

The median equipment repair time method is used when the requirement is defined in terms of an equipment repair time median and is based on log-normally distributed corrective maintenance task times. The mean maintenance time and maximum maintenance time method is used to demonstrate indexes such as mean corrective maintenance time and mean preventive maintenance time.

The mean method is used when the requirement is stated in terms of a mean value and there is a corresponding value of design goal. The preventive maintenance times method is used when the specified index involves the mean preventive maintenance task time and/or maximum preventive maintenance task time at any percentile. The critical percentile method is used when the requirement is expressed in terms of a required critical percentile and a corresponding value of design goal.

The percentiles and maintenance method utilizes a test of proportion for demonstrating fulfillment of maximum preventive maintenance task time at any percentile, median corrective maintenance task time, median preventive maintenance task time, and maximum corrective maintenance task time at the 95th percentile, when preventive and corrective maintenance repair time statistical distributions are not known.

The critical maintenance time method is used when the requirement is stated in terms of required critical maintenance time and a corresponding design-goal value. The man-hour rate method (using simulated faults) is used to demonstrate man-hour rates per operating hour and is based on the predicted equipment failure rate, total accumulative simulated demonstration operating hours, and total accumulative chargeable maintenance man-hours.

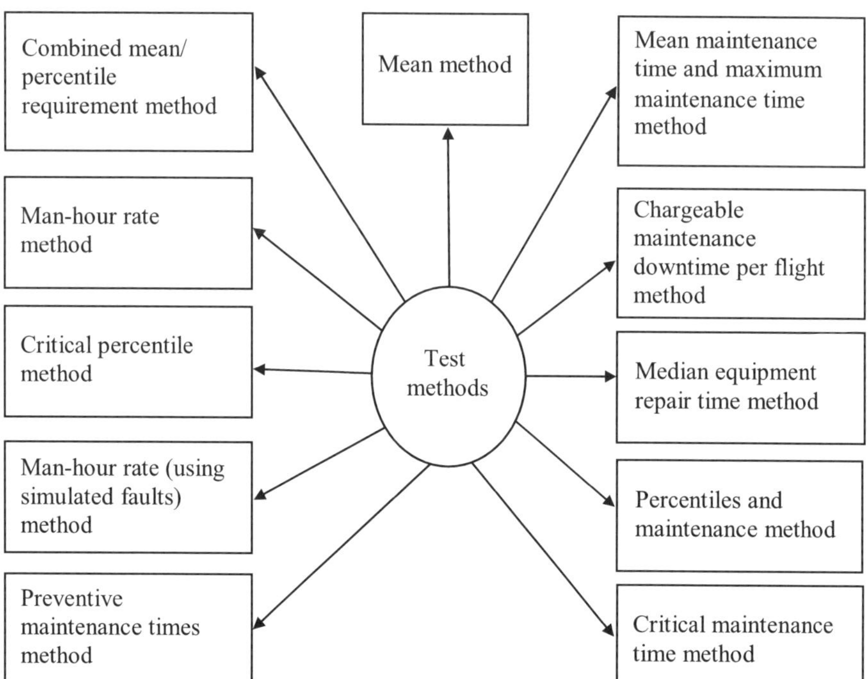

Fig. 7.2 Test methods for demonstrating diverse maintainability parameters

The combined mean/percentile requirement method is used in situations where the requirement is a dual requirement for the mean and for either the 95th or 90th percentile of maintenance times, when maintenance times are log-normally distributed. The chargeable maintenance downtime per flight method is used in testing aircraft; it makes use of the central limit theorem. Furthermore, the chargeable downtime per flight is the allowable time, expressed in hours, to perform the maintenance activity.

The man-hour rate method is used to demonstrate man-hour rates, particularly man-hours per flight hour. Finally, the preventive maintenance time method is used in situations when the specified index involves the mean preventive maintenance task time and/or maximum preventive maintenance task time at any percentile, as well as when all possible preventive maintenance tasks need to be completed.

Information on statistical aspects of many of the above methods is available in Ref. [16].

7.10 Useful Guidelines for Avoiding Pitfalls in Maintainability Testing of Mining Equipment

Some of the useful guidelines to avoid pitfalls in maintainability testing of mining equipment are as follows [18]:

- Tailor MIL-STD-471 [16] for the program and equipment under consideration with care and avoid relying on it totally.
- Aim to perform some "dry run" testing.
- Improve the technical manual verification and validation process well before the maintainability demonstration test.
- Define, rectify, and verify discovered shortcomings/deficiencies and the associated requirements for corrective action.
- Carry out a new and different trial for each and every trial that identifies a shortcoming or deficiency.
- Limit the permissible trial repetitions as a requirement for canceling the test, moving into an "evaluate and fix" phase, and then repeating the complete test with newly stated faults.

7.11 Problems

1. Discuss the following two reliability test classifications:
    - Qualification and acceptance testing
    - Operational testing
2. What is success testing?

3. A manufacturer of a part used in a piece of mining equipment is required to demonstrate 98% reliability of that part at a 95% confidence level. Determine the number of parts to be tested when only zero failures are allowed.
4. What are the ways to perform accelerated testing?
5. A sample of 30 identical engineering items used in mines was tested to failure at a temperature of 120 °C, and their mean time between failures (*MTBF*) was 3000 h. Calculate the *MTBF* of these items at the normal operating temperature of 80 °C.
6. List the five most important documents used, directly or indirectly, in reliability testing.
7. Discuss the following items in regard to maintainability testing:
   - Functional tests
   - Marginal tests
   - Static tests
8. Discuss at least seven test methods for demonstrating diverse maintainability parameters.
9. Discuss the four most useful guidelines for avoiding pitfalls in maintainability testing.
10. Compare reliability testing with maintainability testing.

References

1. AMC Pamphlet AMCP 702-3, Quality Assurance, US Army Material Command (AMC), Washington, DC (1968)
2. Dhillon, B.S.: Design Reliability: Fundamentals and Applications. CRC, Boca Raton, FL (1999)
3. MIL-STD-470, Maintainability Program Requirements, US Department of Defense, Washington, DC (1966)
4. Von Alven, W.H. (ed.): Reliability Engineering. Prentice-Hall, Englewood Cliffs, NJ (1964)
5. Grant Ireson, W., Coombs, C.F., Moss, R.Y.: Handbook of Reliability Engineering and Management. McGraw-Hill, New York (1996)
6. Nelson, W.: Accelerated Testing. Wiley, New York (1980)
7. Meeker, W.Q., Hahn, G.J.: How to Plan an Accelerated Life Test: Some Practical Guidelines. American Society for Quality Control (ASQC), Milwaukee, WI (1985)
8. Bain, L.J., Engelhardt, M.: Statistical Analysis of Reliability and Life Testing Models: Theory. Dekker, New York (1991)
9. Jensen, F., Peterson, N.E.: Burn-In. Wiley, New York (1982)
10. Elsayed, A.E.: Reliability Engineering. Addison Wesley Longman, Reading, MA (1996)
11. Dhillon, B.S.: Medical Device Reliability and Associated Areas. CRC, Boca Raton, FL (2000)
12. Fries, R.C.: Reliability Design of Medical Devices. Dekker, New York (1997)
13. Tobias, P.A., Trindade, D.: Applied Reliability. Van Nostrand Reinhold, New York (1986)
14. AMCP-706-133, Maintainability Engineering Theory and Practice, US Department of Defense, Washington, DC (1976)
15. Dhillon, B.S.: Engineering Maintainability. Gulf, Houston, TX (1999)
16. MIL-STD-471, Maintainability Verification/Demonstration/Evaluation, US Department of Defense, Washington, DC (1966)

17. PRIM-1, A Primer for DOD Reliability, Maintainability and Safety Standards, Reliability Analysis Center, Rome Air Development Center, Griffiss Air Force Base, Rome, New York (1988)
18. Bentz, R.W.: Pitfalls to Avoid in Maintainability Testing. In: Proceedings of the Annual Reliability and Maintainability Symposium, pp. 278–282 (1982)

Chapter 8
Mining Equipment Maintenance

8.1 Introduction

Each year around the globe billions of dollars are spent on equipment maintenance, and since the Industrial Revolution maintenance of engineering equipment has been a challenging issue. Although over the years impressive progress has been made in maintaining engineering equipment in the field environment effectively, it is still a challenge due to factors such as complexity, size, competition, cost, and safety. For example, in the US underground coal mining industry, maintenance of mining equipment accounts for over 30% of the total lost-time injuries [1].

Furthermore, in regard to cost, past experience shows that mining equipment maintenance costs range from around 20% to over 35% of total mine operating costs and are increasing steadily [1]. To control these costs, mining companies have centered their efforts on areas such as optimizing scheduled maintenance operations, deferring nonessential maintenance, reducing maintenance manpower, controlling inventories of spare parts more effectively, and using contract maintenance support [2].

This chapter presents various important aspects of mining equipment maintenance.

8.2 Maintenance-related Facts and Figures

Some of the facts and figures concerned, directly or indirectly, with mining equipment maintenance are as follows:

- Over 25% of the accidents in underground coal mining occur during maintenance activity [3].
- According to injury data for independent contractor employees in the mining industry [4], during the period 1983–1990 about 20% of coal mine injuries occurred during machine maintenance or while using hand tools.

- Equipment maintenance costs range from 20% to over 35% of total mine operating costs [1].
- As per Ref. [5], the total annual cost of engineering maintenance is approx. $450 million in the Australian underground coal mining industry.
- Lubricants account for roughly 1% of maintenance costs for underground coal mining equipment [5].
- In open-pit mines, in both Chile and Indonesia, maintenance cost is greater than 60% of the operating cost [6].
- Usually, the cost of maintenance in the mining industry varies from 40 to 50% of the equipment operating cost [7].
- A multinational Finnish company reported that the direct cost of maintenance in their mines is approx. 33% of the total production cost [7, 8].
- Approximately 10% of production time is lost by unplanned maintenance in Australian underground coal mining industry [5].
- According to various civilian and military studies, it is possible to reduce preventive maintenance and corrective maintenance task times by 40% to 70% with planned maintainability design efforts [1].

8.3 Maintenance Engineering Objectives, Total Productive Maintenance, and Reasons for Its Performance

Maintenance engineering is an analytical, methodical, and deliberate function whose main contributing objectives are as follows [9, 10]:

- Improve the maintenance organization.
- Improve maintenance operations.
- Reduce the maintenance skills required.
- Improve and ensure maximum use of maintenance facilities.
- Reduce the amount of supply support.
- Reduce the effect of complexity.
- Establish the optimum frequency and extent of preventive maintenance to be performed.
- Reduce the frequency and amount of maintenance.

Total productive maintenance (TPM) may simply be described as productive maintenance performed by all personnel through small and autonomous group activities. More specifically, the complete TPM definition incorporates the following elements of development [11]:

- TPM involves each and every employee (*i.e.*, from the managing director to the caretaker on the shop floor).
- TPM is implemented by all departments (*i.e.*, engineering, operation, and maintenance departments).
- TPM aims to maximize the effectiveness of all involved equipment.

- TPM establishes a thorough system of productive maintenance for the entire life span of equipment. This includes daily maintenance (*e.g.*, adjustments, lubrication, and cleaning) and the scheduled inspection of equipment using both predictive and diagnostic methods and allows for the periodic early treatment of problems through timely and preventive repairs.
- TPM is based on the promotion of productive maintenance through the organizational setup as well as on the participation of small autonomous groups of employees who are completely committed to the program success.

Three principal reasons, shown in Fig. 8.1, for the performance of TPM are concerned with eliminating losses that are considered formidable obstacles to equipment effectiveness [11].

Downtime losses are concerned with equipment failure due to breakdowns and setup and adjustment shutdowns to exchange machine tools/dies in manufacturing processes. Defect-related losses are concerned with the following two items:

- Reduction in yield due to problems associated with the time and production losses between equipment startup to a stable production level.
- Process defects due to poor quality production that cause rejects, leading to scrap in the manufacturing process.

Finally, speed-related losses are concerned with the following:

- Reduction in speeds due to differences between designed and real-time speeds (*e.g.*, reduction in flow rates in hydraulic machine or pump transfer systems).
- Idling and minor stoppages due to abnormal operation of sensors, resulting in machine shutdowns or blockages of conveyors, chutes, etc.

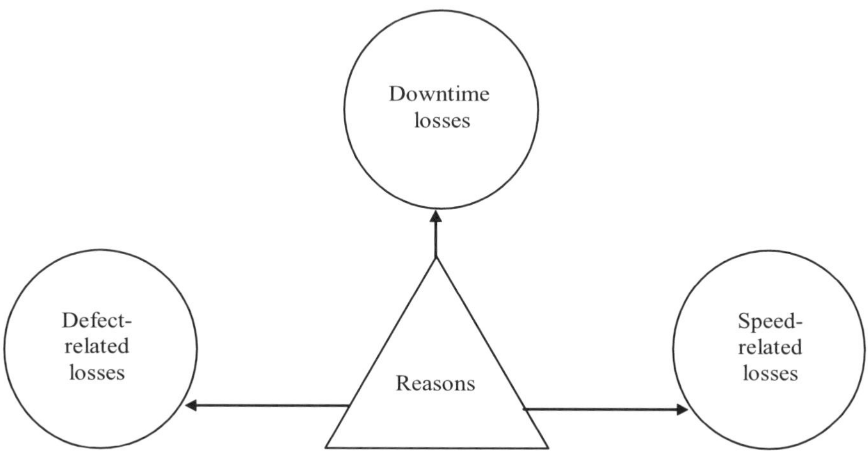

Fig. 8.1 Principal reasons for the performance of total productive maintenance (TPM)

8.4 Factors Contributing to Equipment Maintenance Cost in Mines

Many factors contribute to the cost of mining equipment maintenance. The major ones are shown in Fig. 8.2 [1].

In the case of equipment design, it is important to note that certain makes and models of mining equipment are designed satisfactorily to facilitate maintenance and repair activity, while the basic design of other models hinders maintenance and repair activities. In the case of management attitude toward maintenance, attitudes range from "when it fails, fix it" to strong upper-level management support for effectively planned and implemented preventive maintenance policies and procedures aimed at minimizing unscheduled equipment downtime and controling maintenance costs.

In the case of the skill of maintenance management personnel, the skills needed to manage and organize a mine maintenance program effectively differ quite significantly from the skills required to carry out "hands-on" maintenance of mining

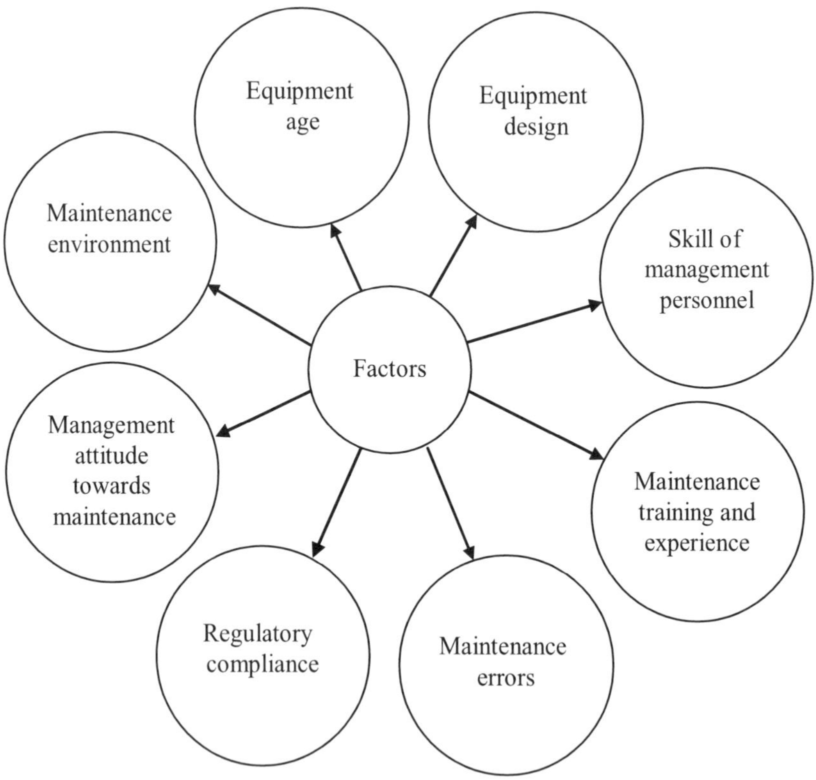

Fig. 8.2 Major factors contributing to mining equipment maintenance costs

equipment. This basically means that poor maintenance management contributes to increased maintenance costs. Regarding equipment age, older equipment tends to be smaller and inherently simpler in design. Consequently, the maintenance of older machines or equipment is somewhat simpler. In contrast, modern equipment/machines tend to be more complex, larger, and overlaid with many "add-on" systems and components, making the basic maintenance process and accessibility rather more difficult.

With respect to the maintenance environment, it is a totally different task to maintain a continuous miner in a, say, 91-cm coal seam than to maintain one in a fairly well-equipped standing-height underground repair workshop. In the case of regulatory compliance, environmental control and safety devices needed to ensure regulatory compliance add to the complexity factor and increase maintenance-related costs.

In the case of maintenance training and experience, poor maintenance skills on the part of maintainers result from factors such as poor training, lack of manuals, guides, and job performance aids, and complexity of maintenance tasks. Finally, in the case of maintenance errors, as per the opinions of involved maintenance manpower, errors such as removing and replacing nonfailed items, failing to install or repair a component correctly, troubleshooting one system too long, failing to test a component prior to reassembly, and not replacing suspected components during a previous opportunity account for approx. 10 to 25% of all maintenance time [1], which results in a substantial increase in maintenance costs.

8.5 Maintenance of Explosion-protected Switchgear in Mines

The maintenance of explosion-protected equipment is a complex subject and is an ongoing process. It may simply be described as routine actions performed to ensure that the equipment remains serviceable and continues to operate within the designed parameters. Thus, in order to achieve the maximum serviceable lifetime of the explosion-protected equipment, it is essential to plan maintenance activity with care. Nonetheless, the maintenance of explosion-protected mining equipment can be divided into the following three areas [12]:

- Visual inspections
- Physical inspections
- Routine work

Visual inspections include checking fasteners (*i.e.*, to see if they are in place), checking labels (*i.e.*, to see if they are in place and legible), checking covers (*i.e.*, to see if they are in place), checking inspection windows (*i.e.*, for cracks, imperfections, etc.), checking operational pushbuttons and handles (*i.e.*, to see if they are operational and not impeded in any manner), checking instruments and indication equipment (*i.e.*, to see if they are in proper working order), checking enclosures (*i.e.*, to make sure the integrity of enclosures is not compromised by any physical damage), and ensuring that the equipment is generally serviceable and not covered in layers of coal dust.

Physical inspections include items such as fasteners (*i.e.*, to make sure they are of the correct type and length and are tightened up), operating handles and pushbuttons (*i.e.*, to make sure they are operating properly and any adjustments are performed and tightened), blank plugs/covers (*i.e.*, to verify that they are in place and fitted correctly), threaded holes (*i.e.*, to make sure threads are not damaged and holes are not full of dirt and fluff), equipment operation (*i.e.*, to verify proper operation of equipment using test equipment), flame paths (*i.e.*, to ensure flame paths are within the correct tolerance and are not pitted or damaged), wiring and major electrical equipment (*i.e.*, to ensure main connections are tight, that there is no evidence of overheating or arcing, and that control wiring is in place and clear of power wiring), operation of protection equipment (*i.e.*, to make sure the equipment operates properly with the integral test buttons on the equipment), and equipment condition (*i.e.*, to make sure there is no physical damage to the equipment and the general condition of equipment is good).

Routine work is concerned with tasks such as thoroughly cleaning and lubricating fasteners, cleaning and lubricating threaded holes, cleaning and lubricating flame paths to inhibit rust and oxidization, checking flame-proof plugs and sockets to make sure they are in good condition and within specified tolerances, cleaning and tightening all control wiring connections, checking to ensure that protection and control relays are operating within the specified parameters and are not damaged or overheated, checking indication equipment and instruments to see if they are operational and working within operational parameters and specifications, cleaning, checking, and tightening power connections and wiring, checking to see if neutral earthing resistors are fitted correctly and connections are tight and that there is no evidence of heat damage, making sure forwarded/reverse isolators are in an operational state and all connections are tightened, ensuring circuit breakers are operational and all auxiliaries, undervoltage releases, and shunt trips are working properly, checking to see if all contactors are operational and within limits according to contact wear gauges, and making sure core balance transformers are in good working condition [11].

8.6 Useful Maintenance Measures for Mines

Usually management utilizes various types of approaches to measuring the effectiveness of the maintenance activity in the organization. Often it employs various indexes to manage and control the maintenance as there is no single index that can effectively reflect the overall performance of the maintenance activity within the organization. The main objective of these indexes is to encourage maintenance personnel to make appropriate improvements over past performance.

This section presents a number of indexes (*i.e.*, broad and specific) considered useful for application in the mining industry [12–16]. The broad indexes basically indicate the overall performance of the organization in regard to the maintenance activity, whereas the specific indexes give the performance in particular areas of maintenance activity. The values of these indexes can be plotted on a periodic basis to show trends.

Index I

This is a broad index and is expressed by

$$\theta_1 = \frac{C_m}{\alpha}, \tag{8.1}$$

where

θ_1 is the index parameter,
C_m is the total maintenance cost,
α is the total output expressed in tons, gallons, etc.

This index is considered quite useful for relating an organization's total maintenance cost to its total output.

Index II

This is a specific index and is used to measure the effectiveness of the maintenance activity. The index is defined by

$$\theta_2 = \frac{DT_b}{DT_t}, \tag{8.2}$$

where

θ_2 is the index parameter,
DT_t is the total downtime,
DT_b is the total downtime caused by breakdowns.

Index III

This index belongs to a broad category and is expressed by

$$\theta_3 = \frac{C_m}{S_t}, \tag{8.3}$$

where

θ_3 is the index parameter,
S_t is the total sales.

According to some studies, average maintenance cost for the entire industry is approximately 5% of sales. However, there is wide variation within the industrial sector.

Index IV

This is a specific index and is defined by

$$\theta_4 = \frac{T_{pm}}{T_{em}}, \tag{8.4}$$

where

θ_4 is the index parameter,
T_{em} is the total time spent for the entire maintenance activity,
T_{pm} is the total time spent for preventive maintenance only.

This index has proven to be a quite useful tool for controling preventive maintenance within a maintenance organization. According to some studies, its value should be kept to within 20% and 40% limits [16].

Index V

This index also belongs to a specific category and is expressed by

$$\theta_5 = \frac{N_j}{I_t}, \tag{8.5}$$

where

θ_5 is the index parameter,
I_t is the total number of inspections completed,
N_j is the number of jobs resulting from inspections.

Index VI

This is a broad index that relates the total maintenance cost to the total investment in plant and equipment. The index is defined by

$$\theta_6 = \frac{C_m}{IN_t}, \tag{8.6}$$

where

θ_6 is the index parameter,
IN_t is the value of the total investment in plant and equipment.

Past experience indicates that average values for θ_6 in the chemical and steel industries are approximately 3.8% and 8.6%, respectively [16].

Index VII

This index belongs to a specific category and is used to measure maintenance effectiveness. The index is expressed by

$$\theta_7 = \frac{MH_\mathrm{e}}{MH_\mathrm{t}}, \tag{8.7}$$

where

θ_7 is the index parameter,
MH_t is the total number of maintenance man-hours worked,
MH_e is the total number of man-hours of emergency and unscheduled jobs.

Index VIII

This is a specific index that relates maintenance costs to manufacturing costs. It is defined by

$$\theta_8 = \frac{C_\mathrm{m}}{MC_\mathrm{t}}, \tag{8.8}$$

where

θ_8 is the index parameter,
MC_t is the total manufacturing cost.

8.7 Mathematical Models for Performing Mining Equipment Maintenance

Over the years, a large number of mathematical models have been developed for application in maintenance activity. This section presents some of the models considered useful for performing mining equipment maintenance [16].

Model I

Inspections are usually performed to reduce equipment downtime, but they are often disruptive. This model is concerned with obtaining the optimum number of inspections per facility per unit time. The model defines the total facility downtime as follows [17, 18]:

$$DT = xT_i + \frac{\mathrm{K}T_\mathrm{f}}{x}, \tag{8.9}$$

where

K	is a constant associated with a specific facility,
x	is the number of inspections per facility per unit of time,
T_i	is the facility downtime per inspection,
T_f	is the facility downtime per failure or breakdown,
DT	is the total downtime per unit of time for a facility.

Differentiating Eq. (8.9) with respect to x we obtain

$$\frac{\mathrm{d}DT}{\mathrm{d}x} = T_i - \frac{\mathrm{K}T_f}{x^2}\,. \tag{8.10}$$

Setting Eq. (8.10) equal to zero and then rearranging we get

$$x^* = \left[\frac{\mathrm{K}T_f}{T_i}\right]^{1/2}, \tag{8.11}$$

where

x^* is the optimum number of inspections per facility per unit of time.

Substituting Eq. (8.11) into Eq. (8.9) we obtain

$$DT^* = 2\left(\mathrm{K}T_iT_f\right)^{1/2}, \tag{8.12}$$

where

DT^* is the total optimal downtime per unit of time for a facility.

Example 8.1

A coal mining facility was observed over a period of time, and we obtained the following data values:

$$K = 4, \quad T_i = 0.02 \text{ month}, \quad T_f = 0.05 \text{ month}\,.$$

Calculate the optimal number of inspections per month using the above specified data values in Eq. (8.11).

Substituting the given data values into Eq. (8.11) yields

$$x^* = \left[\frac{(4)(0.05)}{0.02}\right]^{1/2} = 3.16 \text{ months}\,.$$

Thus, the optimal number of inspections per month is approx. 3.

Model II

This model is concerned with calculating the effective failure rate of a system when one of its units fails, the corrective maintenance is initiated immediately. The system is composed of n independent and identical units in parallel and in which m units are allowed to fail without system failure. The failed system is not repaired and the unit failure and corrective maintenance rates are constant. In this case, an approximate effective failure rate of $(n-m)$ – out-of-n system is given by [19]

$$\lambda_{(n-m)/n} = \frac{n!\lambda^{m+1}}{(n-m-1)!\mu^m}, \tag{8.13}$$

where

λ is the constant failure rate of a unit,
μ is the constant corrective maintenance rate of that unit,
$\lambda_{(n-m)/n}$ is the system approximate effective failure rate.

Example 8.2

Assume that a mining system is made up of four independent, identical, and active units in parallel. At least two units must work normally for the system to succeed and the repair process starts as soon as a unit fails. The failed system is never repaired.

The unit constant failure and repair rates are 0.002 failures per hour and 0.04 repairs per hour, respectively. Calculate the system approximate effective failure rate.

Substituting the given data values into Eq. (8.13) we get

$$\lambda_{(4-2)/4} = \frac{4!\,(0.002)^{2+1}}{(4-2-1)!(0.004)^2} = 0.00012 \text{ failures/h}.$$

Thus, the mining system approximate effective failure rate is 0.00012 failures/h.

Model III

This model is concerned with determining the reliability and mean time to failure (*MTTF*) of a system subject to periodic maintenance. The following assumptions are associated with the model [20, 21]:

- Periodic maintenance is carried out on the system after every x hours, starting at time zero.
- A failed component/part is replaced with a new and statistically identical one.

For periodic maintenance, the time interval of x hours is expressed as

$$x = jX + T, \quad j = 0, 1, 2, \ldots; \ 0 \leq T < X. \tag{8.14}$$

For $j = 1$ and $T = 0$, the reliability of a redundant system subject to periodic maintenance after every X hours is expressed by

$$R_X\left(x = X\right) = R\left(X\right) . \tag{8.15}$$

For $j = 2$ and $T = 0$, we get

$$R_X\left(x + 2T\right) = \left[R\left(X\right)\right]^2 . \tag{8.16}$$

Note that in this case, the system must operate the first X hours without a failure. It must also operate for another X failure-free hours after the replacement of any part that has experienced a failure.

For $0 < T < X$, another T hours of system failure-free operation is needed. Thus, we write

$$R_X\left(x = 2X + T\right) = \left[R\left(X\right)\right]^2 R(T) . \tag{8.17}$$

In general form Eq. (8.17) is expressed as follows:

$$R_X\left(x = jX + T\right) = \left[R\left(X\right)\right]^j \left[R(T)\right] , \quad \text{for } j = 0, 1, 2, \ldots ;\ 0 \leq T < X . \tag{8.18}$$

The redundant system mean time to failure with periodic maintenance is expressed by

$$MTTF_{\mathrm{m}} = \int_0^\infty R_X(x)\,\mathrm{d}x , \tag{8.19}$$

where

$MTTF_{\mathrm{m}}$ is the redundant system $MTTF$ with the performance of periodic maintenance.

To evaluate Eq. (8.19), we write the integral over the range $0 < x < \infty$ as follows:

$$MTTF_{\mathrm{m}} = \sum_{j=0}^{\infty} \int_{jX}^{(j+1)X} R_X(x)\,\mathrm{d}x . \tag{8.20}$$

It is to be noted that in Eq. (8.20) the integral of Eq. (8.19) is divided into time intervals of length X. Thus, for $x = jX + T$, inserting Eq. (8.18) into Eq. (8.20) we obtain

$$MTTF_{\mathrm{m}} = \sum_{j=0}^{\infty} \int_0^X [R(X)]^j R(T)\,\mathrm{d}T . \tag{8.21}$$

Note that in Eq. (8.21) for $x = jX + T$, $\mathrm{d}x = \mathrm{d}T$ and the limits become 0 and X.

Thus, rearranging Eq. (8.21) we get

$$MTTF_{\mathrm{m}} = \sum_{j=0}^{\infty} [R(X)]^j \int_0^X R(T)\,\mathrm{d}T . \tag{8.22}$$

Since

$$\sum_{j=0}^{\infty}[R(X)]^j = \frac{1}{1-R(X)}\,, \tag{8.23}$$

Eq. (8.22) becomes

$$MTTF_{\mathrm{m}} = \frac{\int_0^X R(T)\,\mathrm{d}T}{1-R(X)}\,. \tag{8.24}$$

Example 8.3

Two independent and identical mining machines form a parallel system. Each machine's time to failure is described by the exponential distribution with the mean value of 300 h and the preventive maintenance is carried out after every 150 h. Calculate the system *MTTF* with and without preventive maintenance.

With the aid of Chap. 3 and using the specified data values, we write down the following equation for the parallel system reliability:

$$R(x) = 2\,\mathrm{e}^{-\frac{x}{300}} - \mathrm{e}^{-\frac{2x}{300}}\,. \tag{8.25}$$

Inserting Eq. (8.25) and the specified data into Eq. (8.24) we get

$$MTTF_{\mathrm{m}} = \frac{\int_0^{150}\left[2\,\mathrm{e}^{-\frac{T}{300}} - \mathrm{e}^{-\frac{2T}{300}}\right]\mathrm{d}T}{1-\left[2\,\mathrm{e}^{-\frac{150}{300}} - \mathrm{e}^{-\frac{2(150)}{300}}\right]} = 912.44\,\mathrm{h}\,.$$

System *MTTF* without preventive maintenance is given by

$$\begin{aligned} MTTF &= \int_0^{\infty} R(x)\,\mathrm{d}x \\ &= \int_0^{\infty}\left[2\,\mathrm{e}^{-\frac{x}{300}} - \mathrm{e}^{-\frac{2x}{300}}\right]\mathrm{d}x \\ &= 450\,\mathrm{h}\,. \end{aligned}$$

Thus, the system *MTTF* with and without preventive maintenance is 912.44 h and 450 h, respectively. This simply means that the performance of preventive maintenance has helped to improve system *MTTF* from 450 h to 912.44 h.

Model IV

This mathematical model represents a mining system that can either be in operating or failed state [21]. The failed system is repaired. The system state-space diagram is shown in Fig. 8.3. The numerals in the circle and the box denote system states.

The following assumptions are associated with the model:

- The repaired mining system is as good as new.
- The system failures are statistically independent.
- System failure and repair rates are constant.

The following symbols are associated with the model:

j is the jth system state; $j = 0$ (mining system operating normally), $j = 1$ (mining system failed),
λ_m is the mining system constant failure rate,
μ_m is the mining system constant repair rate,
$P_j(t)$ is the probability that the mining system is in state j at time t, for $j = 0, 1$.

With the aid of the Markov method described in Chap. 3, we write down the following two differential equations for the Fig. 8.3 diagram [21]:

$$\frac{dP_0(t)}{dt} + \lambda_m P_0(t) = \mu_m P_1(t)\,, \tag{8.26}$$

$$\frac{dP_1(t)}{dt} + \mu_m P_1(t) = \lambda_m P_0(t)\,, \tag{8.27}$$

At time $t = 0$, $P_0(t) = 1$ and $P_1(t) = 0$.

Solving Eqs. (8.26) and (8.27) we obtain

$$P_0(t) = \frac{\mu_m}{\mu_m + \lambda_m} + \frac{\lambda_m}{\mu_m + \lambda_m} e^{-(\mu_m + \lambda_m)t}\,, \tag{8.28}$$

$$P_1(t) = \frac{\lambda_m}{\mu_m + \lambda_m} - \frac{\lambda_m}{\mu_m + \lambda_m} e^{-(\mu_m + \lambda_m)t}\,. \tag{8.29}$$

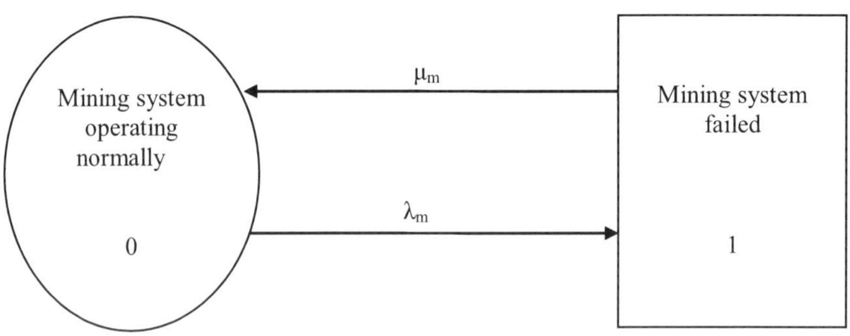

Fig. 8.3 Mining system transition diagram

The mining system availability is given by

$$A_m(t) = P_0(t) = \frac{\mu_m}{\mu_m + \lambda_m} + \frac{\lambda_m}{\mu_m + \lambda_m} e^{-(\mu_m + \lambda_m)t} , \tag{8.30}$$

where

$A_m(t)$ is the mining system availability at time t.

As the value of t becomes very large Eq. (8.30) reduces to

$$A_m = \frac{\mu_m}{\mu_m + \lambda_m} , \tag{8.31}$$

where

A_m is the mining system steady state availability.

Since $\lambda_m = 1/MTTF_m$ and $\mu_m = 1/MTTR_m$, Eq. (8.31) becomes

$$A_m = \frac{MTTF_m}{MTTF_m + MTTR_m} , \tag{8.32}$$

where

$MTTF_m$ is the mining system mean time to failure,
$MTTR_m$ is the mining system mean time to repair.

Example 8.4

Assume that the mean time to failure and mean time to repair of a mining machine are 2000 h and 4 h, respectively. Calculate the machine steady-state availability.

Substituting the given data values into Eq. (8.32) we get

$$A_m = \frac{2000}{2000 + 4} = 0.9980 .$$

Thus, the mining machine's availability is 0.9980. In other words, there is 99.8% probability that the machine will be available for service.

Model V

This mathematical model can be used to represent a mining system that can either fail completely or undergo periodic preventive maintenance (PM). The failed system is repaired. The model is considered quite useful for predicting system availability, the probability of system down for preventive maintenance, and the probability of system failure.

The system state-space diagram is shown in Fig. 8.4. The numerals in the circle and boxes denote system states.

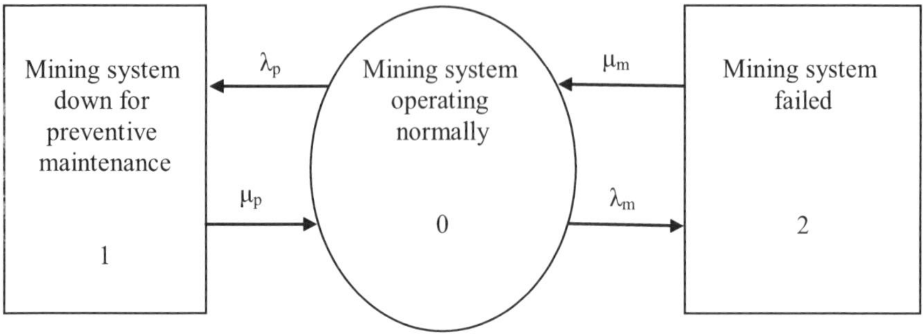

Fig. 8.4 Mining system state-space diagram

The model is subjected to the following assumptions:

- All system failures are statistically independent.
- System failure rate, repair rate, rate of system downtime for preventive maintenance, and rate of system preventive maintenance performance are constant.
- After preventive maintenance or repair, the mining system is as good as new.

The following symbols are associated with the model:

i is the ith system state; $i = 0$ (mining system operating normally), $i = 1$ (mining system down for preventive maintenance), $i = 2$ (mining system failed),

λ_p is the constant rate of the mining system downtime for preventive maintenance,

μ_p is the constant rate of the mining system preventive maintenance performance,

λ_m is the mining system constant failure rate,

μ_m is the mining system constant repair rate,

$P_i(t)$ is the probability that the mining system is in state i at time t; for $i = 0, 1, 2$.

Using the Markov method described in Chap. 3, we write down the following three differential equations for Fig. 8.4 diagram [21, 22]:

$$\frac{dP_0(t)}{dt} + (\lambda_m + \lambda_p) P_0(t) = \mu_m P_2(t) + \mu_p P_1(t), \tag{8.33}$$

$$\frac{dP_1(t)}{dt} + \mu_p P_1(t) = \lambda_p P_0(t), \tag{8.34}$$

$$\frac{dP_2(t)}{dt} + \mu_m P_2(t) = \lambda_m P_0(t). \tag{8.35}$$

At time $t = 0$, $P_0(0) = 1$, $P_1(0) = 0$, and $P_2(0) = 0$.

Solving Eqs. (8.33)–(8.35) we get

$$P_0(t) = \frac{\mu_p\mu_m}{n_1 n_2} + \left[\frac{(n_1+\mu_p)(n_1+\mu_m)}{n_1(n_1-n_2)}\right] e^{n_1 t} - \left[\frac{(n_2+\mu_p)(n_2+\mu_m)}{n_2(n_1-n_2)}\right] e^{n_2 t}, \quad (8.36)$$

$$P_1(t) = \frac{\lambda_p\mu_m}{n_1 n_2} + \left[\frac{\lambda_p n_1+\lambda_p\mu_m}{n_1(n_1-n_2)}\right] e^{n_1 t} - \left[\frac{(\mu_m+n_2)\lambda_p}{n_2(n_1-n_2)}\right] e^{n_2 t}, \quad (8.37)$$

$$P_2(t) = \frac{\lambda_m\mu_p}{n_1 n_2} + \left[\frac{\lambda_m n_1+\lambda_m\mu_p}{n_1(n_1-n_2)}\right] e^{n_1 t} - \left[\frac{(\mu_p+n_2)\lambda_m}{n_2(n_1-n_2)}\right] e^{n_2 t}, \quad (8.38)$$

where

$$n_1, n_2 = \frac{-A \pm \left[A^2 - 4\left(\mu_p\mu_m + \lambda_m\mu_p + \lambda_p\mu_m\right)\right]^{1/2}}{2}, \quad (8.39)$$

$$A = \left(\mu_m + \mu_p + \lambda_m + \lambda_p\right), \quad (8.40)$$

$$n_1 + n_2 = -A, \quad (8.41)$$

$$n_1 n_2 = \mu_p\mu_m + \lambda_p\mu_m + \lambda_m\mu_p. \quad (8.42)$$

The probability of the mining system downtime for preventive maintenance, the probability of the mining system failure, and the mining system availability are given by Eqs. (8.37), (8.38), and (8.36), respectively.

As time t becomes very large, Eq. (8.36) reduces to

$$AV_{mss} = \frac{\mu_m\mu_p}{\mu_p\mu_m + \lambda_p\mu_m + \lambda_m\mu_p}, \quad (8.43)$$

where

AV_{mss} is the mining system steady-state availability.

Example 8.5

Assume that for a mining machine we have the following data values:

$\lambda_m = 0.007$ failures per hour,
$\lambda_p = 0.009$ per hour,
$\mu_m = 0.01$ repairs per hour,
$\mu_p = 0.02$ per hour.

Calculate the mining machine steady-state availability using Eq. (8.43).

Using the specified data values in Eq. (8.43) yields

$$AV_{\mathrm{mss}} = \frac{(0.01)(0.02)}{(0.01)(0.02) + (0.009)(0.01) + (0.007)(0.02)} = 0.4652\,.$$

Thus, the mining machine steady-state availability is 0.4652. In other words, there is an approximately 47% chance that the mining machine will be available for service.

8.8 Problems

1. List at least five facts and figures concerned with mining equipment maintenance.
2. What are the main objectives of maintenance engineering?
3. What is total productive maintenance?
4. List at least five reasons for the performance of total productive maintenance.
5. List at least six major factors that contribute to mining equipment maintenance costs.
6. Discuss the maintenance of explosion-protected switchgear in mines.
7. Prove Eq. (8.12).
8. Assume that the following data values are associated with a mining machine:
   - $\lambda_{\mathrm{m}} = 0.008$ failures/h
   - $\lambda_{\mathrm{p}} = 0.009$ per hour
   - $\mu_{\mathrm{m}} = 0.04$ repairs/h
   - $\mu_{\mathrm{p}} = 0.05$ per hour

   Calculate the mining machine steady state availability using Eq. (8.43).
9. Prove Eq. (8.36).
10. Define three broad maintenance indexes for application in the area of mining.

References

1. Unger, R.L., Conway, K.: Impact of maintainability design on injury rates and maintenance costs for underground mining equipment. In: Improving Safety at Small Underground Mines, compiled by R.H. Peters, Report No. Special Publication 18-94, US Bureau of Mines, Washington, DC, pp. 140–167 (1994)
2. Conway, E.J., Unger, R.: Maintainability Design of Underground Mining Equipment, Report No. USBMOFR 39-91-V1 and V2, US Bureau of Mines (USBM), Washington, DC (1994)
3. Hamilton, D.D., Hopper, J.E., Jones, J.H.: Inherently Safe Mining Systems: Executive Summary, Report No. USBM OFR 124-77, US Bureau of Mines, Washington, DC (1977)
4. Rethi, L.L., Barrett, E.A.: A Summary of Injury Data for Independent Contractor Employees in the Mining Industry from 1983–1990, Report No. USBM IC 9344, US Bureau of Mines, Washington, DC (1993)

5. Clark, D.: Tribology – its application to equipment reliability and maintainability design in the underground coal mining industry. In: Proceedings of the Institution of Engineers Australia Tribology Conference, pp. 38–44 (1990)
6. Hall, R.A., Danshmend, L.K., Lipsett, M.G., Wong, J.: Reliability analysis as a tool for surface mining equipment evaluation and selection. CIM Bull. 93, 1044, 78–82 (2000)
7. Bo Forsman, Kumar, U.: Surface mining equipment and maintenance trends in the Scandinavian countries. J. Mines Minerals Fuels August/September, 266–269 (1992)
8. Harjunpaa, H.: How to determine the effectiveness of an industrial maintenance organization and its influence on the profitability of a company. In: Proceedings of the Euro Maintenance Conference, pp. 101–104 (1992)
9. AMCP 706 – 132, Engineering Design Handbook: Maintenance Engineering Techniques, US Department of Army, Washington, DC (1975)
10. Dhillon, B.S.: Engineering Maintenance: A Modern Approach. CRC, Boca Raton, FL (2002)
11. Leugner, T.: Developing a total productive maintenance (TPM) programme. Quarry Manage. August, 21–25 (1996)
12. Niebel, B.W.: Engineering Maintenance Management. Dekker, New York (1994)
13. Westerkamp, T.A.: Maintenance Manager's Standard Manual. Prentice-Hall, Paramus, NJ (1997)
14. Stoneham, D.: The Maintenance Management and Technology Handbook. Elsevier, Oxford, UK (1998)
15. Hartmann, E., Knapp, D.J., Johnstone, J.J., Ward, K.G.: How to Manage Maintenance. American Management Association, New York (1994)
16. Dhillon, B.S.: Engineering Maintenance: A Modern Approach. CRC, Boca Raton, FL (2002)
17. Wild, R.: Essentials of Production and Operations Management. Holt, Reinhart, and Winston, London (1985)
18. Dhillon, B.S.: Mechanical Reliability: Theory, Models, and Applications. American Institute of Aeronautics and Astronautics, Washington, DC (1988)
19. RADC Reliability Engineer's Toolkit, prepared by the Systems Reliability and Engineering Division, Rome Air Development Center, Griffiss Air Force Base, Rome, NY (1988)
20. Rosenheim, D.E.: Analysis of Reliability Improvement Through Redundancy. In: Proceedings of the New York University Conference on Reliability Theory, pp. 119–142 (June 1958)
21. Von Alven, W.J.: Reliability Engineering. Prentice-Hall, Englewood Cliffs, NJ (1964)
22. Dhillon, B.S.: Reliability Engineering in Systems Design and Operation. Van Nostrand Reinhold, New York (1983)

Chapter 9
Mining Equipment Costing

9.1 Introduction

Cost is an important factor in mining equipment reliability, maintainability, and safety, and over the years it has been increasing at a significant rate. Like any other engineering equipment, cost estimation in mining equipment is extremely important for making effective decisions. The main elements of the total cost of producing a typical piece of mining equipment are design and development cost, materials cost, testing cost, manufacturing cost, operation cost, and maintenance cost.

In particular, based on past experience, the cost of maintaining equipment in the field often varies from 2 to 20 times the original procurement cost [1, 2]. Maintenance cost includes lost opportunities in up-time, yield, rate, and quality due to unsatisfactorily or nonoperating equipment, in addition to the cost associated with equipment-related degradation of safety in regard to people, property, and environment. Needless to say, past experience indicates that 60 to 70% of the projected life cycle cost of a piece of equipment can sometimes be locked in by the end of the preliminary design phase.

This chapter presents various aspects of costing, directly or indirectly, concerned with mining equipment.

9.2 Reasons for Mining Equipment Costing and Methods for Making Mining Equipment Investment Decisions

Some of the important reasons for mining equipment costing are as follows [3, 4]:

- To determine the appropriate selling price of mining equipment.
- To perform new equipment feasibility analyses.
- To ascertain the most profitable approach and materials during equipment manufacture.

- To provide appropriate assistance in controling the cost of equipment manufacture.
- To determine if it is more economical to produce the parts in-house or to procure them from vendors.
- To determine the amount of money required for facilities to manufacture the equipment under consideration.
- To determine the efficiency of the equipment manufacturing process.
- To provide inputs to the long-term financial goals of the organization.

Some of the useful methods that can be used to make mining equipment investment decisions are as follows [5–8]:

- Benefit/cost analysis method
- Return on investment method
- Payback period method

Each of the above methods is presented below, separately.

9.2.1 Benefit/Cost Analysis Method

This method is quite useful in determining if the benefits from the mining equipment project under consideration outweigh its cost. More specifically, the equipment project is considered for development only if its benefits outweigh the investment cost. The benefit/cost ratio is expressed by

$$R_{\mathrm{bc}} = \frac{UB}{IC}\,, \tag{9.1}$$

where

R_{bc} is the benefit-cost ratio,
IC is the investment cost including the operating cost,
UB is the user benefits.

Example 9.1

Assume that a piece of mining equipment is being considered for acquisition, and its estimated cost and expected useful life are \$500,000 and 15 years, respectively. The total benefits of the equipment, in dollars, over its lifespan are estimated to be around \$900,000. Calculate the benefit-cost ratio if the annual maintenance cost of the equipment is \$2000.

Thus, the total investment cost of the mining equipment is

$$IC = 500{,}000 + (2000)(15) = \$530{,}000\,.$$

Substituting the above calculated value and the other given data into Eq. (9.1) we get

$$R_{bc} = \frac{900{,}000}{530{,}000} = 1.67\,.$$

Thus, the value of the benefit-cost ratio is 1.67.

9.2.2 Return on Investment Method

This is another method that can also be used to make mining equipment investment decisions; it uses the following two different approaches to calculate the return on investment.

Approach I

In this case, the return on average investment is calculated by using the following equation:

$$RI_a = \frac{ANP(100)}{(0.5I_a + C_w)}\,, \tag{9.2}$$

where

RI_a is the return on investment,
ANP is the annual average net profit,
I_a is the total amount invested,
C_w is the working capital.

Approach II

In this case, the return on original investment is calculated by using the following equation:

$$RI_o = \frac{ANP(100)}{(I_a + C_w)}\,, \tag{9.3}$$

where

RI_o is the return on the original investment.

Example 9.2

A coal mining organization is considering purchasing a piece of equipment to be used in mines that requires $500,000 in investment and $25,000 in working capital. It is estimated that the use of the equipment will generate average net profits of $30,000 per year.

Calculate the return on the original investment.

Substituting the specified data values into Eq. (9.3) yields

$$RI_o = \frac{(30{,}000)(100)}{500{,}000 + 25{,}000} = 5.71\% \,.$$

Thus, the return on the original investment is 5.71%.

9.2.3 Payback Period Method

This method is concerned with determining the payback period over which the project capital expenditures can be recovered. The major drawback of the method is that it does not consider the profit factor after the recovery of capital expenditures.

The method makes use of the following four approaches to compute the payback period [4, 7].

Approach I

In this case, the payback period is expressed by

$$PP_1 = \frac{TCE}{ACF} \,, \tag{9.4}$$

where

PP_1 is the payback period for approach I,
TCE is the total capital expenditure,
ACF is the average annual cash flow.

The average annual cash flow is expressed by

$$ACF = AP_n + (TCE)(CDR) \,, \tag{9.5}$$

where

AP_n is the average annual net profit,
CDR is the invested capital (fixed) depreciation rate.

Approach II

In this case, the payback period is expressed by

$$PP_2 = \frac{TCE}{AGP} \,, \tag{9.6}$$

where

PP_2 is the payback period for approach II,
AGP is the average annual gross profit.

Approach III

In this case, the payback period is expressed by

$$PP_3 = \frac{1}{[CDR + \{\theta/(1 - ITR)\}]}, \tag{9.7}$$

where

PP_3 is the payback period for approach III,
ITR is the income tax rate,
θ is the minimum acceptable rate of return on investment.

Approach IV

In this case, the payback period is expressed by

$$PP_4 = \frac{TCE}{AP_n}, \tag{9.8}$$

where

PP_4 is the payback period for approach IV.

The average annual net profit is given by

$$AP_n = AGP - [ITR\{AGP - (CDR)(TCE)\} + i(TCE + C_w)], \tag{9.9}$$

where

i is the interest rate on the borrowed money,
C_w is the working capital.

Example 9.3

A mining organization is considering replacing a piece of equipment used in underground operations, at a cost of \$400,000. It is estimated that the use of the equipment will generate an average saving of \$30,000 per year. Calculate the capital expenditure recovery period.

Using the specified data values in Eq. (9.5) yields

$$PP_2 = \frac{400{,}000}{30{,}000} = 13.3 \text{ years}.$$

Thus, the capital expenditure recovery period is 13.3 years.

9.3 Cost Estimation Models for Mining Equipment

Over the years, many mathematical models have been developed to estimate various types of costs concerning engineering equipment [9,10]. This section presents some of these models considered useful for application to mining equipment.

9.3.1 Cost-capacity Model

This model is quite useful for obtaining quick cost estimates for similar new projects, facilities, or equipment of different capacities. The cost of the new item is expressed as follows [10]:

$$C_{\mathrm{n}} = C_{\mathrm{so}} \left[\frac{CP_{\mathrm{n}}}{CP_{\mathrm{so}}} \right]^{\alpha} , \tag{9.10}$$

where

C_{n} is the cost of the new project, facility, or equipment;
C_{so} is the cost of the old but similar project, facility, or piece of equipment;
CP_{n} is the capacity of the new project, facility, or piece of equipment;
CP_{so} is the capacity of the old but similar project, facility, or piece of equipment;
α is the cost-capacity factor whose value varies for different items or projects. For example, the proposed values of this factor for pumps, tasks, and heaters are 0.6, 0.7, and 0.8, respectively [11–13]. In the event of having no available data for the factor, it is considered quite reasonable to assume its value as 0.6.

Example 9.4

Assume that a mining organization spent a total of $500 million to construct a mining site whose annual output is 20,000 tonnes. The organization is considering constructing another similar site whose annual output will be 30,000 tonnes. Estimate the cost of the new site if the value of the cost-capacity factor is 0.6.

Substituting the given data values into Eq. (9.10) we get

$$C_{\mathrm{n}} = (500) \left[\frac{30{,}000}{20{,}000} \right]^{0.6} = \$637.71\,\text{million} .$$

Thus, the cost estimate for the new mining site is $637.71 million.

9.3.2 Corrective Maintenance Labor Cost Estimation Model

This model is concerned with estimating the annual labor cost of corrective maintenance associated with equipment with known mean time between failures and mean time to repair. The equipment annual corrective maintenance labor cost is defined by

$$CMLC = (ASOH)\,(R)\left[\frac{MTTR}{MTBF}\right], \tag{9.11}$$

where

$CMLC$	is the equipment annual corrective maintenance labor cost,
R	is the hourly corrective maintenance labor cost rate,
$ASOH$	is the equipment annual scheduled operating hours,
$MTBF$	is the equipment mean time between failures,
$MTTR$	is the equipment mean time to repair.

Example 9.5

Assume that a piece of mining equipment is scheduled for 6000 h of operation in one year and its estimated mean time between failures and mean time to repair are 2000 h and 10 h, respectively. Calculate the equipment annual corrective maintenance labor cost if the hourly maintenance labor cost rate is $40.

Substituting the given data values into Eq. (9.11) we obtain

$$CMLC = (6000)(40)\left[\frac{10}{2000}\right] = \$1200\,.$$

Thus, the mining equipment annual corrective maintenance labor cost is $1200.

9.3.3 Total Maintenance Labor Cost Estimation Model

This model is concerned with estimating total maintenance labor cost. The total maintenance labor cost is expressed by [10]

$$MLC_{\mathrm{t}} = m\,(AH_{\mathrm{t}})\,(R_{\ell})\,(1+\beta)\,, \tag{9.12}$$

where

MLC_{t}	is the total maintenance labor cost,
R_{ℓ}	is the labor rate per hour,
m	is the total number of employees,
AH_{t}	is the total number of annual hours associated with the maintenance activity,
β	is the benefit ratio.

Example 9.6

Assume that the following data values are known for the maintenance department of a mining company:

- $R_\ell = \$40$,
- $m = 15$ employees,
- $AH_t = 2000\,\text{h}$,
- $\beta = 0.4$.

Calculate the company's total maintenance labor cost.

Substituting the given data values into Eq. (9.12) we get

$$MLC_t = 15(2000)(40)(1+0.4) = \$1{,}680{,}000\,.$$

Thus, the mining company's total maintenance labor cost is \$1,680,000.

9.3.4 Production Facility Downtime Cost Estimation Model

This model is concerned with estimating the production facility downtime cost. The production facility downtime cost is expressed by [14]

$$PFDC = C_r + C_{r\ell} + C_o + C_p + C_{ru} + C_{ti}\,, \tag{9.13}$$

where

$PFDC$	is the production facility downtime cost,
C_r	is the ruined product replacement cost,
$C_{r\ell}$	is the revenue-loss cost,
C_o	is the idle operator salary,
C_p	is the late-penalty cost,
C_{ru}	is the replacement unit rental cost (if any),
C_{ti}	is the tangible and intangible costs associated with factors such as loss of good will and customer dissatisfaction.

9.3.5 Motor Operation Cost Estimation Model

This model is used to estimate the cost of operating an alternating-current (AC) motor. Thus, the motor operating cost is expressed by

$$MAOC = \frac{(0.746)\,(MAOH)\,(EC)\,(MS)}{\alpha}\,, \tag{9.14}$$

where

$MAOC$ is the motor annual operating cost,
α is the motor efficiency,
MS is the motor size expressed in horsepower,
EC is the electricity cost expressed in dollars per kilowatt hour ($/KWH),
$MAOH$ is the motor annual operating hours.

Example 9.7

Assume that a 60-horsepower AC motor is operated for 3000 h per year and the cost of electricity is $0.15/KWH. The motor efficiency is 95%.

Calculate the motor annual operating cost.

Inserting the specified data values into Eq. (9.14) we get

$$MAOC = \frac{(0.746)(3000)(0.15)(60)}{0.95} = \$21{,}202.1\,.$$

Thus, the motor annual operating cost is $21,202.10.

9.3.6 Failure Mode and Effect Analysis Cost Estimation Model

This model is concerned with estimating the cost of performing Failure Mode and Effect Analysis (FMEA). The cost of performing FMEA is expressed by

$$C_{\mathrm{f}} = nC_{\mathrm{h}}\,, \tag{9.15}$$

where

C_{f} is the cost of performing FMEA,
n is the number of hours needed to perform FMEA,
C_{h} is the hourly cost of performing FMEA.

The number of hours needed to perform FMEA is expressed by

$$n = m(17.79), \tag{9.16}$$

where

m is the number of unique items requiring FMEA, the number of circuit cards for piece part and circuit-level FMEA, or the number of pieces of equipment for equipment level FMEA. The value of m varies from 3 to 206 [9].

9.3.7 Reliability Testing Cost Estimation Model

This model is concerned with estimating the cost of reliability testing. Thus, the cost of reliability testing is expressed by [9]

$$RTC = (RTC_{\text{h}})\,(T)\ , \tag{9.17}$$

where

RTC	is the reliability testing cost,
T	is the number of hours needed to perform reliability testing,
RTC_{h}	is the hourly cost of performing reliability testing.

The number of hours needed to perform reliability testing is expressed by

$$T = (182.07)\,\theta\ , \tag{9.18}$$

where

θ is the factor whose value depends on the complexity of hardware; for more than 25,000 parts ($\theta = 3$), for between 15,000 and 25,000 parts ($\theta = 2$), and for less than 15,000 parts ($\theta = 1$).

9.4 Life Cycle Costing Concept

This is a widely used concept in the industrial sector; its history may be traced back to 1965, when the Logistics Management Institute of Washington, D.C. prepared a document entitled "Life Cycle Costing in Equipment Procurement" [15]. The life cycle cost of an item may be described as the sum of all costs (*i.e.*, acquisition and ownership) over its entire life span.

Some of the important reasons for the increasing use of the life cycle costing concept in the industrial sector are competition, rising maintenance costs, budget limitations, costly products (*e.g.*, military systems, aircraft), greater ownership costs in comparison to acquisition costs, and increasing cost-effectiveness awareness among product users. The life cycle costing concept can be used for various purposes. Some of the important ones are as follows [15, 16]:

- Selecting among options
- Forecasting future budget needs
- Deciding the replacement of aging equipment
- Choosing the most beneficial procurement policy
- Making strategic decisions and design tradeoffs
- Determining cost drivers
- Improving comprehension of basic design-associated parameters in product design and development

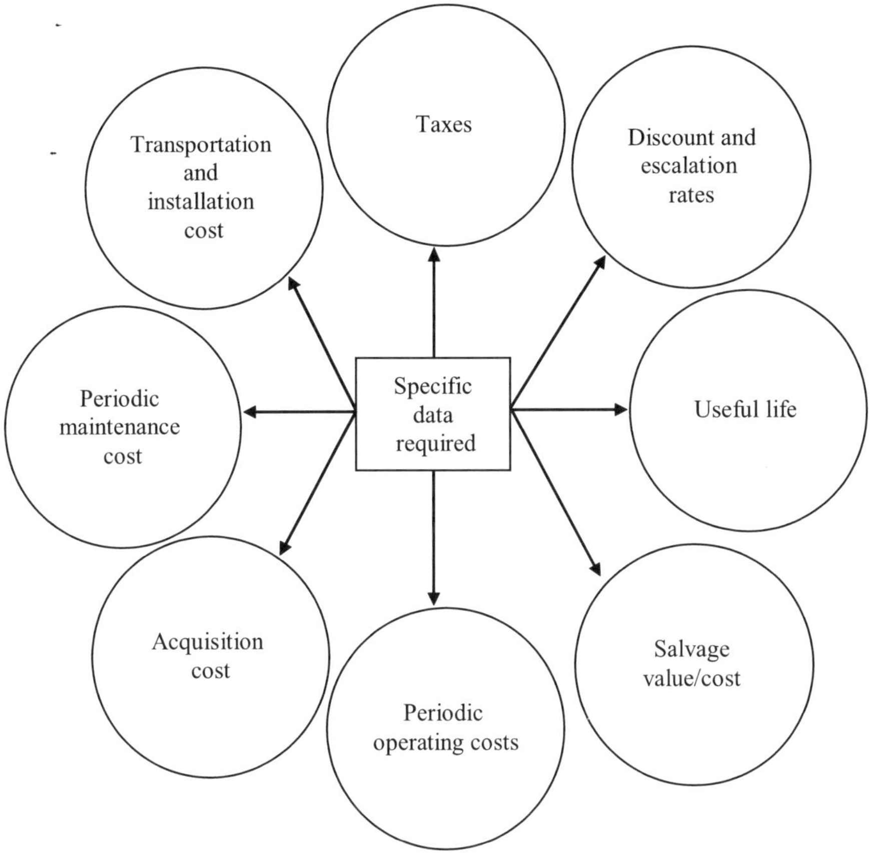

Fig. 9.1 Specific data required to perform life cycle costing studies of equipment

Various types of data are required to perform life cycle costing studies of equipment. Some of the specific data required are shown in Fig. 9.1 [10, 15–17].

9.5 Life Cycle Costing Steps

A life cycle costing study can be performed by following seven steps listed below [18].

- Determine useful life of the item/equipment under consideration.
- Obtain estimates of all involved costs (*e.g.*, acquisition and maintenance costs).
- Determine item/equipment terminal cost.
- Subtract the item/equipment terminal cost from its ownership cost.
- Discount the end result of the preceding step to the present value.

- Calculate the item/equipment life cycle cost by adding the result of the previous step to the item/equipment procurement cost.
- Repeat the above steps for all items/equipment under consideration for acquisition.

9.6 Life Cycle Cost Estimation Models for Mining Equipment

Over the years, many mathematical models have been developed to estimate the life cycle cost of items/equipment [10]. They may be classified under two categories: general and specific. The general life cycle cost models are not tied to any specific item, whereas the specific life cycle cost models are tied to specific items/equipment. This section presents two general models considered useful for application in the mining industry.

9.6.1 Model I

In this case, the item/equipment life cycle cost is classified under two major categories: recurring cost and nonrecurring cost. Thus, the item/equipment life cycle cost is expressed by

$$LCC = C_{\mathrm{r}} + C_{\mathrm{n}} , \tag{9.19}$$

where

LCC	is the item/equipment life cycle cost,
C_{r}	is the item/equipment recurring cost,
C_{n}	is the item/equipment nonrecurring cost.

The item/equipment recurring cost, C_{r}, is

$$C_{\mathrm{r}} = \sum_{i=1}^{5} RC_i , \tag{9.20}$$

where

RC_i is the ith recurring cost: ($i = 1$) operating, ($i = 2$) maintenance, ($i = 3$) labor, ($i = 4$) inventory, and ($i = 5$) support.

Similarly, the item/equipment nonrecurring cost, C_{n}, is expressed by

$$C_{\mathrm{n}} = \sum_{i=1}^{10} NRC_i , \tag{9.21}$$

where

NRC_i is the ith nonrecurring cost: ($i = 1$) acquisition, ($i = 2$) transportation, ($i = 3$) research and development, ($i = 4$) training, ($i = 5$) support, ($i = 6$) life cycle cost management, ($i = 7$) installation, ($i = 8$) equipment qualification approval, ($i = 9$) reliability and maintainability improvement, and ($i = 10$) test equipment.

9.6.2 Model II

In this model, the item/equipment life cycle cost is divided into four areas: research and development cost, production and construction cost, operation and support cost, and disposal cost. Thus, mathematically the item/equipment life cycle cost is expressed by

$$LCC = C_{\mathrm{rd}} + C_{\mathrm{pc}} + C_{\mathrm{os}} + C_{\mathrm{d}}\ , \tag{9.22}$$

where

C_{rd} is the item/equipment research and development cost,
C_{pc} is the item/equipment production and construction cost,
C_{os} is the item/equipment operation and support cost,
C_{d} is the item/equipment disposal cost.

The item/equipment research and development cost is made up of seven major components: research cost, product planning cost, engineering design cost, software cost, test and evaluation cost, design documentation cost, and life cycle management cost.

The main components of the item/equipment production and construction cost are construction cost, quality control cost, industrial engineering and operations analysis cost, manufacturing cost, and initial logistics support cost. Three major elements of the item/equipment operation and support cost are product distribution cost, sustaining logistic support cost, and product/equipment operation cost.

Finally, the item/equipment disposal cost is expressed by

$$C_{\mathrm{d}} = C_{\mathrm{er}} + k\theta\left(C_{\mathrm{od}} - V_{\mathrm{r}}\right)\ , \tag{9.23}$$

where

C_{er} is the equipment ultimate retirement cost,
θ is the condemnation factor,
k is the total number of unscheduled maintenance actions,
V_{r} is the reclamation value,
C_{od} is the overall equipment disposal cost.

Example 9.8

A mining organization using a certain piece of equipment in its underground operations is contemplating replacing it with a better one. Three manufacturers are bidding to sell the equipment. The data given in Table 9.1 are available on these three manufacturers' equipment.

Using life cycle costs, determine which of the three pieces of equipment the mining organization would purchase.

Using the Table 9.1 data, we get the following annual expected failure costs of equipment A, B, and C, respectively:

$$EFC_{\mathrm{A}} = (0.05)(8000) = \$400\,,$$
$$EFC_{\mathrm{B}} = (0.06)(9000) = \$540\,,$$

and

$$EFC_{\mathrm{C}} = (0.04)(7000) = \$280\,,$$

where

EFC_{A} is the annual expected failure cost of equipment A,
EFC_{B} is the annual expected failure cost of equipment B,
EFC_{C} is the annual expected failure cost of equipment C.

Using the given and calculated data values and Ref. [10], we get the following present values of equipment A, B, and C failure costs over their useful lives:

$$PVEFC_{\mathrm{A}} = 400\left[\frac{1-(1+0.07)^{-10}}{0.07}\right] = \$2809.4\,,$$

$$PVEFC_{\mathrm{B}} = 540\left[\frac{1-(1+0.07)^{-10}}{0.07}\right] = \$3792.7\,,$$

and

$$PVEFC_{\mathrm{C}} = 280\left[\frac{1-(1+0.07)^{-10}}{0.07}\right] = \$1966.6\,,$$

Table 9.1 Data for three pieces of mining equipment manufactured by three different manufacturers

No.	Description	Equipment A	Equipment B	Equipment C
1	Selling price	\$400,000	\$350,000	\$425,000
2	Useful operating life	10 years	10 years	10 years
3	Annual failure rate	0.05 failures	0.06 failures	0.04 failures
4	Cost of a failure	\$8000	\$9000	\$7000
5	Annual operating cost	\$6000	\$8000	\$4000
6	Annual interest rate	7%	7%	7%

where

$PVEFC_{\mathrm{A}}$	is the present value of equipment A operating costs over its useful life,
$PVEFC_{\mathrm{B}}$	is the present value of equipment B operating costs over its useful life,
$PVEFC_{\mathrm{C}}$	is the present value of equipment C operating costs over its useful life.

Using the given and calculated values, the life cycle costs of equipment A, B, and C are

$$\begin{aligned} LCC_{\mathrm{A}} &= 400{,}000 + PVEFC_{\mathrm{A}} + PVEOC_{\mathrm{A}} \\ &= 400{,}000 + 2809.4 + 42{,}141.5 \\ &= \$444{,}950.9\,, \end{aligned}$$

$$\begin{aligned} LCC_{\mathrm{B}} &= 350{,}000 + PVEFC_{\mathrm{B}} + PVEOC_{\mathrm{B}} \\ &= 350{,}000 + 3792.7 + 56{,}188.7 \\ &= \$409{,}981.4\,, \end{aligned}$$

and

$$\begin{aligned} LCC_{\mathrm{C}} &= 425{,}000 + PVEFC_{\mathrm{C}} + PVEOC_{\mathrm{C}} \\ &= 425{,}000 + 1966.6 + 28{,}094.3 \\ &= \$455{,}060.9\,, \end{aligned}$$

where

LCC_{A}	is the life cycle cost of equipment A,
LCC_{B}	is the life cycle cost of equipment B,
LCC_{C}	is the life cycle cost of equipment C.

As the life cycle cost of equipment B is the lowest, it should be purchased by the mining organization.

9.7 Problems

1. List at least seven important reasons for mining equipment costing.
2. Define the term "life cycle cost".
3. Outline the steps to perform a life cycle costing study.
4. List at least eight specific data items required to perform life cycle costing studies of, say, mining equipment.
5. What is the benefit/cost analysis method?
6. A mining organization is considering purchasing a piece of equipment to be used in mines that requires \$600,000 in investment and \$40,000 in working capital. It is estimated that the use of the equipment will generate average net profits of \$27,000 per year.
   Calculate the return on the original investment.

Table 9.2 Data for mining equipment under consideration

No.	Description	Manufacturer A's equipment	Manufacturer B's equipment
1	Selling price	\$300,000	\$350,000
2	Annual failure rate	0.04 failures	0.05 failures
3	Useful operating life	15 years	15 years
4	Cost of a failure	\$4000	\$3500
5	Annual operating cost	\$10,000	\$9000
6	Annual interest rate	6%	6%

7. Describe payback period method.
8. Assume that a piece of mining equipment is scheduled for 5000 h of operation in one year and its estimated mean time between failures and mean time to repair are 2500 h and 5 h, respectively. Calculate the equipment annual corrective maintenance labor cost if the hourly maintenance labor cost rate is \$30.
9. A mining organization spent \$400 million to construct a mining site whose annual output is 15,000 tonnes. The organization is considering constructing another similar site whose annual output will be 35,000 tonnes. Estimate the cost of the new site if the value of the cost-capacity factor is 0.8.
10. A mining organization is considering purchasing a piece of mining equipment, and two manufacturers are bidding to sell the equipment. Table 9.2 presents data available on both the manufacturers' equipment. Determine which of the two pieces of equipment the mining organization would purchase based on their life cycle costs.

References

1. Dhillon, B.S.: Engineering Maintenance: A Modern Approach. CRC, Boca Raton, FL (2002)
2. Niebel, B.W.: Engineering Maintenance Management. Dekker, New York (1994)
3. Doyle, L.E.: How to Estimate Costs of New Products. In: Chironis, N.P. (ed.) Management Guide for Engineers and Technical Administrators. McGraw-Hill, New York, pp. 115–126 (1969)
4. Dhillon, B.S.: Medical Device Reliability and Associated Areas. CRC, Boca Raton, FL (2000)
5. Baker, N.R., Pound, W.J.: R and D project selection: where we stand. IEEE Trans. Eng. Manage. 11, 124–134 (1964)
6. Clarke, T.E.: Decision making in technologically based organizations: a literature survey of present practice. IEEE Trans. Eng. Manage. 21, 9–23 (1974)
7. Kasner, E.: Essentials of Engineering Economics. McGraw-Hill, New York (1979)
8. Cetron, M.J., Martino, J., Roepcke, L.: The selection of R and D program content: survey of quantitative methods. IEEE Trans. Eng. Manage. 14, 4–13 (1967)
9. RADC Reliability Engineer's Toolkit, prepared by the Systems Reliability and Engineering Division, Rome Air Development Center (RADC), Griffiss Air Force Base, Rome, NY (1988)
10. Dhillon, B.S.: Life Cycle Costing. Gordon and Breach, New York (1989)
11. Desai, M.B.: Preliminary cost estimating of process plants. Chem. Eng. 65–70 (July 1981)

12. Jelen, F.C., Black, J.H. (eds.): Cost and Optimization Engineering. McGraw-Hill, New York (1983)
13. Dieter, G.E.: Engineering Design. McGraw-Hill, New York (1983)
14. Lavitt, J.: The Handbook of Maintenance Management. Industrial, New York (1997)
15. Seldon, M.R.: Life Cycle Costing: A Better Method of Government Procurement. Westview, Boulder, CO (1979)
16. Lamar, W.E.: Technical Evaluation Report on Design to Cost and Life Cycle Cost, North Atlantic Treaty Organization (NATO) Advisory Group for Aerospace Research and Development (AGARD), Advisory Report No. 165, (May 1981). Available from the National Technical Information Service (NTIS), Springfield, VA
17. Brown, R.J.: A new market tool: life cycle costing. Ind. Market Manage. 8, 109–113 (1979)
18. Coe, C.K.: Life cycle costing by state governments. Public Administrat. Rev. September/October, pp. 564–569 (1981)

Chapter 10
Introduction to Mining Equipment Safety

10.1 Introduction

Mine safety is an important issue that is all too frequently overlooked in an industrial sector that has been squeezed hard by recent economic events. A mining accident is quite dangerous, and often deadly accidents occur during the process of mining various types of minerals from underneath the Earth's surface. Some of the causes for the occurrence of mining accidents are dust explosions, flooding, asphyxiant gases, and general mechanical errors from improperly used or malfunctioning engineering equipment.

In the past, underground mining equipment basically consisted of simple but rugged machines powered by hydraulics and electric motors. Today such equipment has become powerful, complex, and sophisticated and requires a high degree of skill to operate and maintain it economically, effectively, and safely. For example, a series of accidents, including two traffic accidents involving a private car, a minibus, and an excavator, at a major copper mine in Chile, halted production for 23 h and cost the company around $700,000 in lost production [1].

In 1977, the US Congress passed the Mine Safety and Health Act (*i.e.*, Mine Act) to improve safety in the nation's mines. As a result of this Act, the US Department of Labor established an agency called the Mine Safety and Health Administration (MSHA) [2]. The main goal of this agency is to enforce compliance with mandatory safety and health standards as a means to eliminate fatal accidents, to minimize the severity and frequency of nonfatal accidents, to reduce health-related hazards, and to promote better health and safety conditions in US mines.

This chapter presents various important aspects of mining equipment safety.

10.2 Facts and Figures

Some of the facts and figures concerning, directly or indirectly, mining equipment safety are as follows:

- In 2006, in the USA 72 miners were killed at work, 47 in coal mining [3, 4].
- Annual mining deaths in the USA decreased from around 500 in the late 1950s to about 93 during the 1990s [5].
- During the period 1991–1999, an average of 21,351 mine-related injuries per year occurred in the USA [5].
- During the period 1983–1992, a total of 81 fatalities occurred in quarries in the UK [6].
- In 2004, 17% of the 37,445 injuries in US underground coal mines were connected with bolting machines [7].
- During the period 1990–1999, 197 equipment fires caused 76 injuries in coal mining operations in the USA [8].
- As per Ref. [9], over 25% of underground coal mining accidents occurred during maintenance activity.
- During the period 1983–1990, according to injury data for independent contractor employees in the mining industry, around 20% of the coal-mine-related injuries occurred during machine maintenance activity or while using hand tools [10].
- During the period 1978–1988, maintenance activity accounted for 34% of all lost-time injuries in US mines [11].
- The average fatality rate in the US mining industry was 27.7 per 100,000 workers, as opposed to 4.8 per 100,000 workers in all US industries during the period 1992–2001 [12].
- During the period 1990–1999, electricity was the fourth leading cause of deaths in the US mining industry [13].

10.3 Quarry Accidents and Electrical-, Equipment Fire-, and Maintenance-related Mining Accidents

Nowadays, quarries are considerably more automated and working conditions in them are more comfortable and safer than decades ago. Unfortunately, many fatalities still occur in quarries. For example, according to a UK study, during the period 1983–1993, 81 fatalities occurred in UK quarries [6]. The breakdown of these fatalities is as follows [6]:

- Vehicle-related (*i.e.*, run over by a vehicle, trapped under vehicle body, vehicles colliding with plant or other vehicles, vehicle overturned on quarry floor or road, and vehicles running over open edge of quarry face bench or ramp): 41%
- Stumbling, falling, and slipping: 13%
- Trapped between belt and head/tail drum rollers: 11%

- Falling objects or ground: 8%
- Maintenance (*i.e.*, while maintaining plant or equipment): 5%
- Crusher blockage (*i.e.*, while clearing feeder or crusher blockages): 4%
- Buried in material: 4%
- Contact with overhead electricity lines: 3%
- Miscellaneous: 11%

A detailed analysis of these fatal accidents revealed that human error was an important factor in areas such as vehicles; maintenance; stumbling, falling, and slipping; and clearing blockages in feeders, crushers, or conveyors.

During the period 1990–1999, a total of 1926 electrical accidents occurred in US mines. Seventy-five of these accidents were fatal [13]. This means electricity was the fourth leading cause of deaths in US mines during the specified period. Furthermore, about 50% of all mine electrical injuries and fatalities occur during the electrical maintenance activity.

A detailed analysis of the data revealed that core problem areas were circuit breakers (*i.e.*, 313 of 1926 accidents) cables (*i.e.*, 309 of 1926 accidents), batteries (*i.e.*, 242 of 1926 accidents), working on energized circuits (*i.e.*, 163 of 1926 accidents), grounding (*i.e.*, 204 of 1926 accidents), and using meters and test leads for troubleshooting (*i.e.*, 90 of 1926 accidents) [13].

Equipment fires can be very dangerous to the safety of miners, particularly when they occur in confined spaces of underground mines. For example, during the period 1990–1999, a total of 197 equipment fires occurred in US coal mines that caused 76 injuries [8]. These fires were analyzed under four categories in a National Institute for Occupational Safety and Health (NIOSH) study. The categories and their corresponding core findings were as follows [8]:

- **Underground equipment fires.** There were a total of 26 equipment fires in underground coal mines, and 10 of them caused 10 injuries.
- **Surface fires at underground mines.** There were a total of 14 equipment fires at the surface of underground coal mines, and 4 of these fires caused 4 injuries.
- **Surface coal mine equipment fires.** There were a total of 140 equipment fires at the surface coal mines, and 56 of them caused 56 injuries.
- **Prep plant fires.** There were a total of 17 equipment fires at the coal preparation plants, and 6 of these fires caused 7 injuries.

Various types of accidents occur during mining equipment maintenance. Some of the commonly occurring maintenance-related accidents in underground coal mines are presented in Table 10.1 [14].

10.4 Causes of Mining Equipment Accidents and Major Ignition Sources for Mining Equipment Fires

Over the years many studies have been performed to identify the causes of mining equipment accidents. One such study was conducted by the US Bureau of Mines

Table 10.1 Some commonly occurring maintenance-related accidents in underground coal mines

No.	Type of accident
1	Falling object
2	Stationary object
3	Flying object
4	Struck by object/machinery
5	Caught, moving-stationary
6	Overexertion, lifting
7	Contact with hot object
8	Rolling object
9	Inhalation of noxious fumes
10	Caught, moving objects
11	Overexertion, push-pull
12	Fall on object

(now NIOSH) [15]. The study revealed seven main causes of mining equipment accidents, as shown in Fig. 10.1 [15]. However, with respect to design another US Bureau of Mines study concluded that no matter how much expertise the engineers had or the type of design problem being considered, typical engineers fail to consider human factors effectively when designing [16].

According to Ref. [8], major ignition sources for equipment fires in US coal mines during the period 1990–1999, in descending order, were as follows:

- Hydraulic fluid/fuel on equipment hot surfaces
- Flame cutting/welding spark/slag
- Electric short/arcing
- Engine malfunction

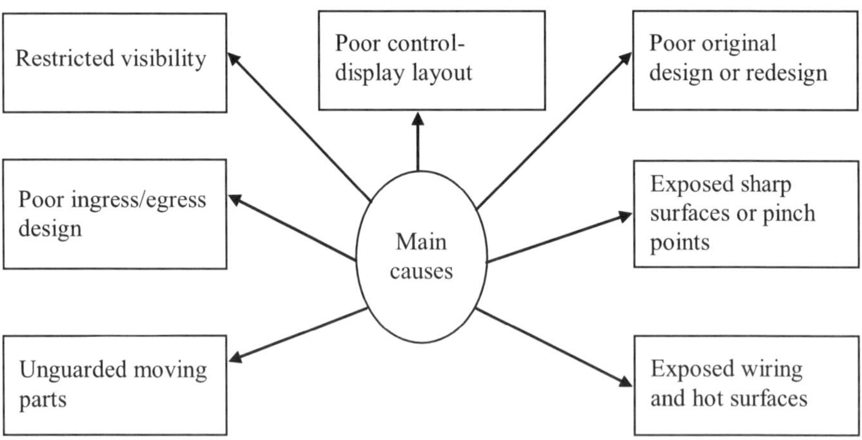

Fig. 10.1 Main causes of mining equipment accidents

10.5 Methods for Performing Mining Equipment Safety Analysis

There are many methods that can be used to perform mining equipment safety analysis [17]. Six of these methods considered most useful to perform mining equipment safety analysis are preliminary hazards analysis, failure modes and effects analysis, management oversight and risk tree (MORT) analysis, consequence analysis, binary matrices, and human reliability analysis [18]. Each of these methods is described below, separately.

10.5.1 Preliminary Hazards Analysis

Preliminary hazards analysis (PHA) is an effective method to identify the potential hazardous conditions inherent in a system and evaluate the significance of potential accidents. The method is composed of ten steps, shown in Fig. 10.2 [18].

Some of the main advantages of PHA are as follows [18]:

- It is a simple and straightforward safety analysis method that qualitatively considers all aspects of a system.
- Anyone possessing a detailed knowledge of the system under consideration can perform an effective PHA.
- It is a fairly cost-effective approach.

In contrast, the disadvantages of PHA include use of a columnar format that may cause the analyst to fall into a "form-filling mode," the fact that it is a qualitative approach, etc.

The application of PHA to a mining system is demonstrated in Ref. [18]. Nonetheless, most of the hazardous energy sources and hazardous processes and events that may be associated with a mining system are presented in Tables 10.2 and 10.3, respectively [18]. This information could be useful in performing PHA of a mining system.

10.5.2 Failure Modes and Effect Analysis (FMEA)

This method is widely used in the industrial sector to perform reliability and safety analyses of engineering systems. FMEA is an effective qualitative approach used to identify all possible ways in which particular pieces of equipment can fail and the possible resulting effects on the total system and associated personnel. Thus, the method can be used to examine the failures of elements of a mining system and determine the effects of such failures on people and objects at the mine. The main steps involved in performing FMEA are as follows [18]:

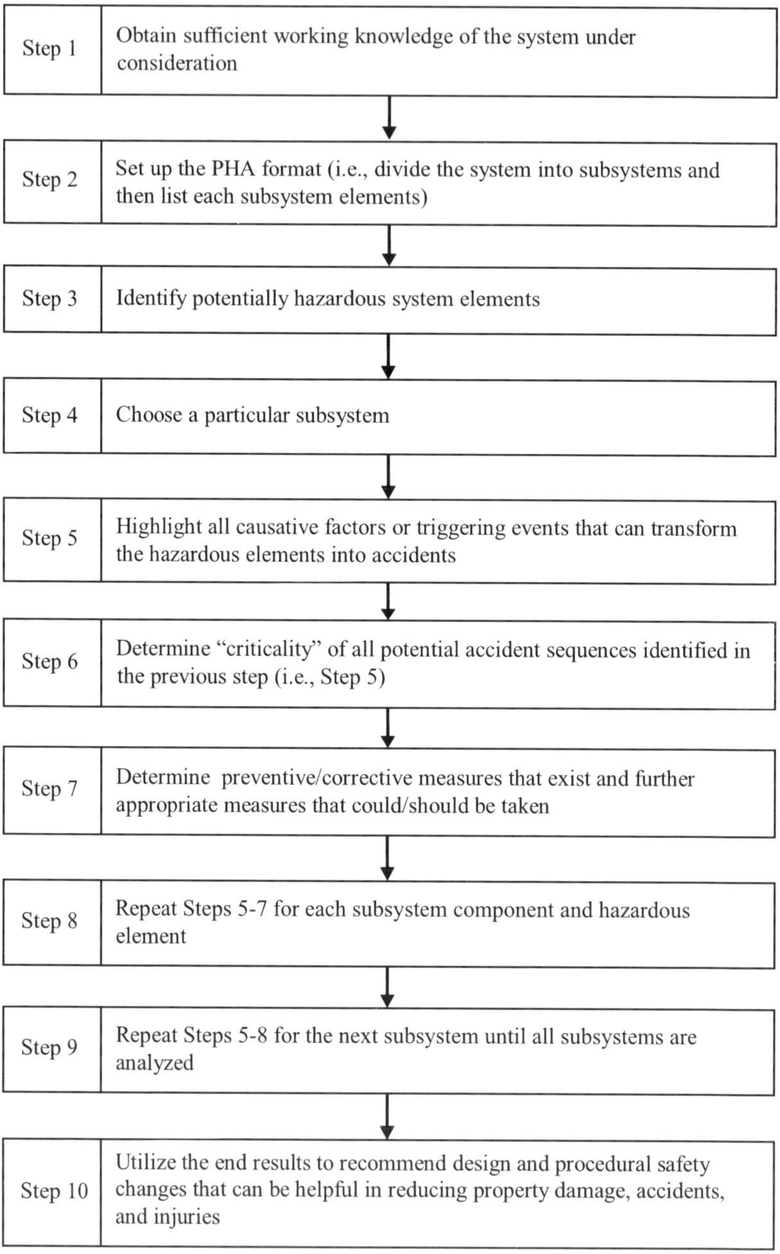

Fig. 10.2 Preliminary hazard analysis (PHA) steps

Table 10.2 Potentially hazardous energy sources in a mining system

No.	Hazardous energy source
1	• Pressure containers
2	• Mechanical equipment
3	• Electrical generators
4	• Rotating machinery
5	• Charged electrical capacitors
6	• Suspension systems
7	• Fuels and mine gases
8	• Storage batteries
9	• Spring-loaded devices
10	• Impacting machinery
11	• Pumps, fans, and blowers
12	• Heating devices

Table 10.3 Potentially hazardous processes and events in a mining system

No.	Hazardous process/event
1	• Mechanical shock
2	• Drilling
3	• Fire
4	• Vibration
5	• Pressure
6	• Explosion
7	• Climbing
8	• Moisture (high humidity)
9	• Acceleration
10	• Heat and temperature
11	• Hoisting (kinetic energy)
12	• Conveying
13	• Excavation

- **Step 1.** Obtain sufficient working knowledge of the system under consideration.
- **Step 2.** Establish the appropriate FMEA format.
- **Step 3.** Choose the first subsystem for analysis and list its components.
- **Step 4.** Choose the first component to be analyzed.
- **Step 5.** Identify and analyze all possible failure modes of the initial component and their anticipated effects.
- **Step 6.** Choose and analyze the remaining components of the first subsystem by repeating step 5 for each component or part.
- **Step 7.** Continue until all subsystems and components have been analyzed by repeating steps 4 to 6 for each subsystem.
- **Step 8.** Prepare a list of items considered critical.

Some of the main advantages of FMEA are that it has a simple analysis procedure, is a useful visibility tool for management, employs a procedure that starts from the detailed level and works upward, and its format provides an orderly and structured examination of the hazardous conditions inherent in an industrial process. In contrast, the major drawback of using FMEA is that it considers only one failure at a time. Additional information on FMEA is available in Chap. 3.

The application of FMEA to an underground coal mining system is demonstrated in Ref. [18].

10.5.3 Management Oversight and Risk Tree (MORT) Analysis

The MORT method is based basically on a document prepared by W.G. Johnson in 1973 [19] and is a comprehensive safety assessment tool that can be applied to any mine safety program. The method focuses on administrative or programmatic control of hazardous conditions and is designed to identify, evaluate, and prevent safety-related errors, omissions, and oversights by workers and management personnel that can cause accidents. The main steps used in performing MORT analysis are shown in Fig. 10.3 [18].

Some of the main advantages of the MORT analysis are that it is a comprehensive approach that attempts to evaluate all aspects of safety in any type of work, is useful for examining all three aspects of an industrial system (*i.e.*, humans, hardware, and management) as they collectively cause accidents, and the results of MORT analysis can suggest appropriate improvements to an existing safety program that could be helpful in saving lives, reducing injuries, and reducing property damage.

Some of the disadvantages of the MORT analysis method are that it creates a vast amount of complex detail, is a time-consuming approach, and emphasizes management's responsibility to provide a safe workplace. The application of the MORT analysis to a mining system is demonstrated in Ref. [18].

10.5.4 Consequence Analysis

Consequence analysis is concerned with determining an undesired event's impact on items such as adjacent property, people, or the environment. Some examples of an undesired event are fire, explosion, projection of debris, and the release of toxic material. The primary consequences of concern in the area of mining include injuries, losses due to property/equipment damage and operational downtime, and fatalities. Needless to say, consequence analysis is one of the intermediate steps of safety analysis as accident consequences are normally determined, initially using approaches such as FMEA or PHA.

Additional information on consequence analysis in regard to its application in the area of mining is available in Ref. [18].

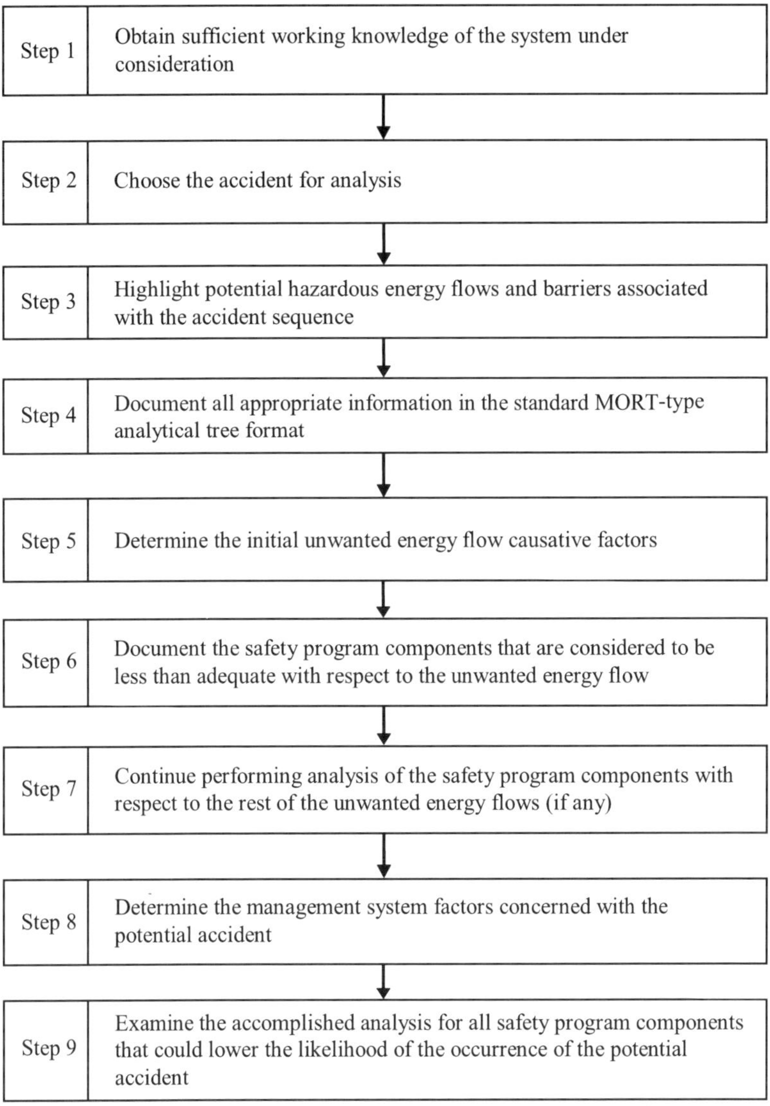

Fig. 10.3 Main steps for performing MORT analysis

10.5.5 Binary Matrices

As many safety analysis methods do not adequately address interactions among the elements or subsystems of a system, in this regard, a binary matrix is a useful, logical, and qualitative approach to identifying system interactions [20]. This method can be used during the system-description stage of safety analysis or as a final checkpoint in a PHA or FMEA to ensure that all important dependencies associated with the system have been considered in the analysis in an effective manner.

The specific tool used in binary matrices is the binary matrix that contains information on the relationships between the elements/components of a system. The objective of this matrix is to identify the one-on-one dependencies that exist between system elements/components. All in all, the matrix serves merely as an effective tool to "remind" the involved analyst that failures in one part of a system may affect, to a certain degree, the normal operation of other, seemingly unrelated, subsystems in completely separate areas.

The application of the method to a mining system is demonstrated in Ref. [18].

10.5.6 Human Reliability Analysis

Human reliability analysis may simply be described as the study of human performance within a complex man-machine operating system. The method is quite useful for developing qualitative information concerning the causes and effects of human error in specific conditions. Human reliability analysis is performed by following the six steps listed below [18].

- **Step 1.** Describe the goals and functions of the system under consideration.
- **Step 2.** Describe all possible situational characteristics.
- **Step 3.** Describe the characteristics of the involved personnel.
- **Step 4.** Describe the tasks and jobs performed by the involved personnel.
- **Step 5.** Perform an analysis of the tasks and jobs to identify error-likely situations and other possible problems.
- **Step 6.** Suggest appropriate changes to the system under consideration.

All of the above steps are described in detail in Refs. [18,21]. The application of the method to a mining system is demonstrated in Ref. [18].

10.6 Hazardous Area Signaling and Ranging Device (HASARD) Proximity Warning System

Over the years a large number of surface and underground mining personnel have been killed or permanently disabled because of working near machinery and powered haulage. For example, in surface mining operations, around 13 people are killed each year by being run over or pinned by mobile mining systems/equipment [22]. In the area of underground mining in the USA, during the period 1988–2000, around 23 fatalities were associated with mining personnel getting caught, pinned, or crushed by continuous mining equipment. An analysis of these fatalities revealed that, although in most cases the mining workers are aware of the dangers, they become preoccupied with operating their equipment and fail to realize when they or their associates stray into or are subjected to hazardous conditions.

To overcome problems such as these an active proximity warning system called HASARD (hazardous area signaling and ranging device) was developed by the National Institute for Occupational Safety and Health (NIOSH). The system warns mine workers when they approach known hazardous areas around heavy mining equipment and other hazardous work zones. HASARD is composed of the following two items [22]:

- **Transmitter:** generates a 60-KHz magnetic field using one or more wire loop antennas. Each antenna is adjusted as the need arises to establish a magnetic field pattern for each hazardous zone.
- **Receiver:** a magnetic field strength meter worn by mining workers. The received signal is compared with preset levels, which are calibrated to highlight danger levels. The receiver outputs can be made to disable machine operations and can include vibratory, audible, and visual indicators.

All in all, HASARD has proven to be a very useful tool and is rugged enough to withstand the harshest production environments. Additional information on HASARD is available from Ref. [22].

10.7 Human-Factor-related Tips for Safer Mining Equipment, Guidelines to Improve Electrical Safety in the Mining Industry, and Strategies to Reduce Mining Equipment Fires and Injuries

As equipment is the primary cause of injury in about 11% of all mining accidents and a secondary cause in another 10%, it is very important that new equipment incorporate good human-factor design criteria that maximize the safety of mine workers [23]. In this regard, human-factor-related tips for safer mining equipment are divided into four areas as follows [23]:

- Control design
- Workstation layout
- Seating
- Visibility

Some of the useful tips concerning control design are as follows: (i) ensure that operators can identify the appropriate controls, quickly and accurately; (ii) ensure that all controls have sufficient resistance to reduce the possibility of inadvertent activation by the weight of a hand or foot; (iii) ensure that all design controls comply with anthropometric data on human operators; (iv) ensure that all design controls can withstand or guard against abuse, such as from falling roof and ribs or from the forces imposed during a panic response in an emergency stop; and (iv) ensure that the speed of a vehicle or component is proportional to the control displacement from its rest position and in the same direction.

Some of the useful tips associated with workstation layout are the following: (i) ensure that the workstation fits operators from the 5th to 95th percentile range; (ii) ensure that the relative placement of controls and displays for similar types of equipment is maintained; (iii) anticipate all potential safety hazards and required emergency actions prior to starting the design process; and (iv) distribute workload as evenly as possible between hands and feet.

Some of the useful tips concerning seating are: (i) ensure that the seat provides appropriate design features for guarding against shocks caused by rough roads and minor collisions that tend to unseat the involved individual; (ii) design the seat in such a way that mine workers can easily maintain or replace it; (iii) ensure that the seat fits and adjusts to body dimensions, supports posture, and distributes weight to relieve pressure points; (iv) ensure that the seat does not hinder the ability of the operator to enter or exit the workstation; and (v) ensure that the seat does not hinder the operator's ability to control the equipment/machine.

Finally, two of the useful tips concerning visibility are: (i) ensure that the workstation provides an unobstructed line of sight to locations or objects that should be visible to carry out a task safely and effectively and (ii) ensure that there is adequate contrast between the luminance of the object or location of interest and the surrounding background, so that the task can be carried out safely and effectively.

Some of the useful guidelines to improve electrical safety in the mining industry are as follows [13]:

- Improve system design.
- Use power line avoidance devices.
- Improve electrical maintenance procedures and schedules.
- Provide power line awareness training.
- Target training at known problem areas.

Finally, some of the strategies to reduce mining equipment fires and injuries are as follows [8]:

- Perform equipment hydraulic, fuel, and electrical system inspections more thoroughly and frequently.
- Develop more rapid equipment/cab fire detection systems having an audible/visible cab alarm.
- Provide frequent emergency preparedness training to individuals operating the equipment.
- Improve systems concerned with equipment/cab fire prevention/suppression.
- Develop appropriate new technologies for fire barriers, emergency hydraulic line drainage/safeguard system, and emergency engine/pump shutoff.
- Develop appropriate and effective local firefighting response capabilities.

10.8 General Areas for Safety Improvements in Mines

Over the years, many general areas for safety improvements in the mining industry have been identified. Most of these areas are shown in Fig. 10.4 [12]. In the area of machinery, improved hardware, operating conditions, and machine design guidelines are considered essential to reduce injuries to mine workers working on or near equipment.

In the area of slips and falls, better procedures and equipment are considered essential to eliminating injuries while mounting or dismounting from a vehicle with ladders. In the area of powered haulage, the development and implementation of new and improved collision warning systems is considered essential to reducing worker exposure to moving vehicles and equipment. Furthermore, better methods to reduce operator jolting and jarring on mobile equipment will be helpful to reduce back injuries.

In the areas of high-wall collapse, new hazard-recognition methods for identifying unstable high walls and slopes are considered necessary so that innovative hazard mitigation strategies can be used effectively. With respect to electrical, the use of overhead power line contact proximity alarms for mobile equipment is consid-

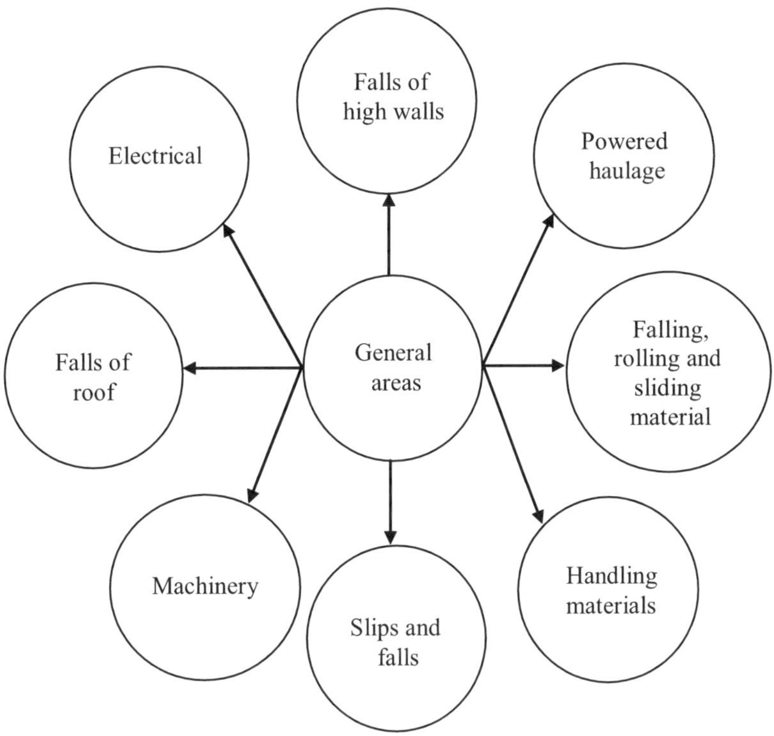

Fig. 10.4 General areas for safety improvements in the mining industry

ered essential on more mobile equipment. In the area of falling, rolling, and sliding material, appropriate methods to detect, mitigate, and/or eliminate material bridging in storage piles, silos, bins, and hoppers need to be developed for reducing injuries from collapsing materials.

As for the collapse of roofs, existing technology to warn of ground-stability-related hazards needs to be used by a larger proportion of the mining sector. Finally, in the area of handling materials, appropriate changes in the way manual tasks are performed or mechanization of such tasks are considered necessary to reduce back-related injuries.

10.9 Problems

1. List at least five facts and figures directly or indirectly concerned with mining equipment safety.
2. Discuss quarry-related accidents.
3. Discuss mining equipment fire-related accidents.
4. List at least ten commonly occurring maintenance-related accidents in underground coal mines.
5. What are the main causes of mining equipment accidents?
6. List and discuss at least four methods that can be used effectively to perform mining equipment safety analysis.
7. Describe the hazardous area signaling and ranging device (HASARD).
8. List at least five useful guidelines to improve electrical safety in the mining industry.
9. Discuss strategies to reduce mining equipment fires and injuries.
10. List and discuss the five most useful general areas for safety improvements in mines.

References

1. Woof, M.: Avoiding Disaster (May 2002). Available online at www.intecvideo.com/AvoidingDisaster.htm
2. Mine Safety and Health Administration (MSHA), US Department of Labor, Washington, DC. Available online at www.msha.gov/contactmsha/contactmsha.htm
3. All Mining Fatalities By State, Mine Safety and Health Administration, US Department of Labor, Washington, DC, January 15, 2007. Available online at www.msha.gov/stats/charts/allstatesnew.asp
4. Coal Fatalities By State, Mine Safety and Health Administration, US Department of Labor, Washington, DC, January 15, 2007. Available online at www.msha.gov/stats/charts/coalbystate.asp
5. Historical Data on Mine Disasters in the United States, US Department of Labor, Washington, DC. Available online at www.msha.gov/MSHAINFO/FactSheets/MSHAFACTS8.htm
6. Scott, A.: Killing off errors. Mine Quarry 14–18 (May 1995)

7. Burgess-Limerick, R., Steiner, L.: Preventing injuries: analysis of injuries highlights high priority hazards associated with underground coal mining equipment. Am. Longwall Mag. 19–20 (August 2006)
8. De Rosa, M.: Equipment fires cause injuries: recent NIOSH study reveals trends for equipment fires at U.S. coal mines. Coal Age 28–31 (October 2004)
9. Hamilton, D.D., Hopper, J.E., Jones, J.H.: Inherently Safe Mining Systems, Report No. USBM OFR 124 -77, US Bureau of Mines (USBM), Washington, DC, (1977)
10. Rethi, L.L., Barett, E.A.: A Summary of Injury Data for Independent Contractor Employees in the Mining Industry from 1983–1990, Report No. USBMIC 9344, US Bureau of Mines, Washington, DC (1983)
11. MSHA Data for 1978–1988, Mine Safety and Health Administration (MSHA), US Department of Labor, Washington, DC
12. Iannacchione, A., Mucho, T.: 100 years of improvement in aggregate worker safety. Stone Sand Gravel Rev. 28–34 (March/April 2003)
13. Cawley, J.C.: Electrical Accidents in the Mining Industry, 1990–1999. IEEE Trans. Ind. Appl. 39, 6, 1570–1576 (2003)
14. Unger, R., Conway, K.: Impact of maintainability design on injury rates and maintenance costs for underground mining equipment. In: Improving Safety at Small Underground Mines, compiled by R.H. Peters, Report No. 18-94, US Bureau of Mines, Department of the Interior, Washington, DC (1994)
15. Sanders, M.S., Shaw, B.E.: Research to Determine the Contribution of System Factors in the Occurrence of Underground Injury Accidents, Report No. US BM OFR 26-89, US Bureau of Mines, Washington, DC (1988)
16. Sanders, M.S., Peay, J.M.: Human Factors in Mining, Report No. USBM IC 9182, US Bureau of Mines, Washington, DC (1988)
17. Dhillon, B.S.: Engineering Safety: Fundamentals, Techniques, and Applications. World Scientific, River Edge, NJ (2003)
18. Daling, P.M., Geffen, C.A.: User's Manual of Safety Assessment Methods for Mine Safety Officials, Report No. BuMines OFR 195 (2) – 83, US Bureau of Mines, Department of the Interior, Washington, DC (1983)
19. Johnson, W.G.: The Management Oversight and Risk Tree-MORT, Report No. SAN 821-2, US Atomic Energy Commission, Washington, DC (1973)
20. Cybulskis, P., *et al.*: Review of Systems Interaction Methodologies, Report No. NUREG/CR-1896, Battelle Columbus Laboratories, Columbus, OH (1981)
21. Swain, A.D., Guttman, H.E.: Handbook of Human Reliability Analysis with Emphasis on Nuclear Power Plant Applications, Report No. NUREG/CR-1278, Sandia National Laboratories, Albuquerque, NM (1982)
22. Hazardous Area Signalling and Ranging Device (HASARD), Pittsburgh Research Laboratory, National Institute for Occupational Safety and Health (NIOSH), Atlanta, GA
23. Unger, R.L.: Tips for safer mining equipment, Holmes Safety Assoc. Bull. 14–15 (October 1996)

Chapter 11
Programmable Electronic Mining System Safety

11.1 Introduction

Generally, mining was traditionally a low-tech industrial sector. Today, it is driven by competitive pressures to go high-tech by using programmable electronics (PE) in areas such as mine processing equipment, mine monitoring systems, longwall mining systems, and automated haulage. Just as in other areas, the application of PE technology (*i.e.*, software, programmable logic controllers (PLCs), and microprocessors) in the mining sector has created unique challenges for system design, verification, operation, maintenance, and assurance of functional safety [1]. More specifically, although the application of PE provides numerous benefits, it certainly adds a level of complexity that, if not considered with care, may compromise the safety of mine workers (*i.e.*, it can create new hazards or worsen the existing ones) [2, 3].

PE technology has various types of unique failure modes that are quite different from mechanical or hard-wired electronic systems traditionally used in the mining sector. Needless to say, the mining industry's experience with the functional safety of PE is quite limited in comparison to other industries. This simply means that the application of the PE technology in the mining sector, with respect to safety, must be considered with care.

This chapter presents various important aspects of PE mining system safety.

11.2 Programmable Electronics Usage Trends in Mining

The mining industry is increasingly using new technologies including programmable electronics (PE) due to various factors. A study of informal industry surveys, industry studies, and published equipment surveys revealed that the usage of PE in the mining sector is not limited to specific systems, mining methods, or commodities [4]. In fact, their usage can be grouped under three basic areas, as shown in Fig. 11.1 [4]. These are control, protection, and monitoring. Within the framework

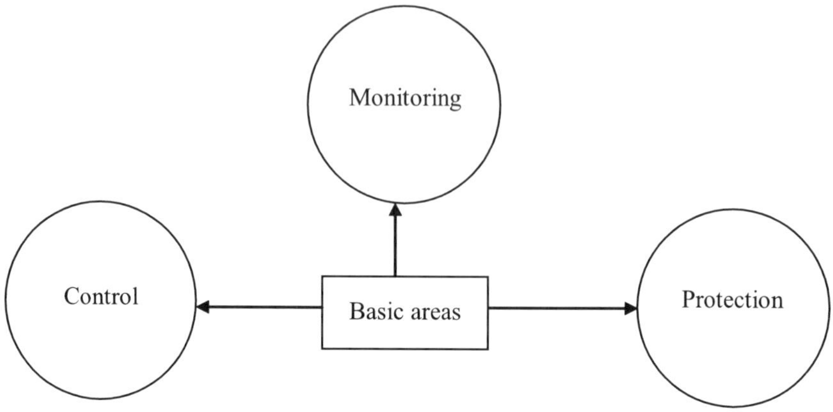

Fig. 11.1 Basic areas of programmable electronics usage

of these three basic areas are many application areas. Some of these are as follows [4]:

- Longwall coal mining systems
- Mine elevators and hoists
- Remote controllers for underground mining machines
- Automated haulage vehicles for surface and underground metal/nonmetal mines
- Mine atmospheric monitoring systems that monitor methane, carbon monoxide, and fresh airflow

Underground mine atmospheric monitoring and control began in the late 1970s, and in the 1990s almost 17% of all underground mines were equipped with computer-based systems [4]. Similarly, from 1990 to 1996, usage of programmable longwall systems doubled, *i.e.*, to around 95% of all longwalls in the USA. Furthermore, microprocessor technology is increasingly being used in the control and monitoring of conveyor systems.

Today, industrywide trends are inclined toward more use and complexity as machinery in general moves from localized PE control to distributed control of processes and machines. All in all, this trend is expected to increase in the future due to factors such as economic pressures, lower grades of ores and coal, and increasing degrees of difficulties in physically accessing such resources.

11.3 Programmable-electronic-related Mishaps

Today programmable-electronic-related mishaps are an important issue in the mining industry. The Mine Safety and Health Administration's (MSHA) serious thinking about the functional safety of PE-based mining systems started in 1990 as a result of the occurrence of an unplanned longwall shield mishap [5]. During the period

1995–2001, a total of 11 PE-related mining incidents were reported in the USA; four of these were fatalities [1,6]. During the same time period (*i.e.*, 1995–2001), 71 incidents were reported in underground coal mines in New South Wales (NSW), Australia [1].

A study of both sets of data, *i.e.*, from the USA and Australia, revealed that the majority of mishaps involved unexpected movements or startups of PE-based mining systems. In 1991, MSHA performed a study of all longwall installations with respect to PE. The study revealed that 35% had experienced unexpected movements basically due to the following problems [1,5]:

- Sticking or defective solenoid valves
- Software programming errors
- Operator error
- Water ingress

A detailed analysis of the above US and Australian data sets reported the four major factors shown in Fig. 11.2, which contributed to PE-based mishaps [5,7]. However, the solenoid valve problems were the leading factor contributing to PE-based mishaps.

In response to the above longwall mishaps, MSHA recommended improvements in items as follows:

- Operator training
- Timely maintenance
- Maintaining alertness for abnormal operational sequences that might be indicative of a software problem
- Maintaining integrity of enclosure sealing

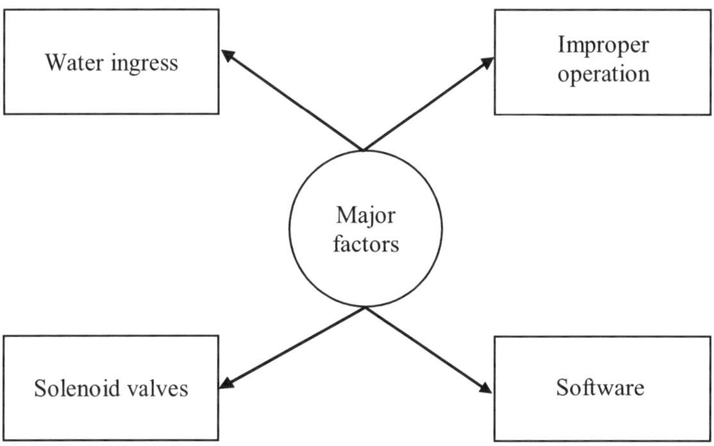

Fig. 11.2 Major factors contributing to programmable-electronics-based mishaps

Subsequently, MSHA realized the shortcomings of the above approach for complex PE mining systems and then proposed a safety framework largely based on the IEC 61508 safety life cycle [8].

11.4 Methods for Performing Hazard and Risk Analysis of Programmable Electronic Mining Systems

Many methods can be used to perform hazard and risk analysis of programmable electronic mining systems. The most useful ones are as follows [9]:

- Preliminary hazard analysis (PHA)
- Failure mode and effect analysis (FMEA)
- Fault tree analysis (FTA)
- Hazard and Operability studies (HAZOP)
- Event tree analysis (ETA)
- Interface analysis
- Action error analysis (AEA)
- Operating and support analysis (OASA)
- Sequentially timed events plot (STEP) investigation system
- Potential or predictive human error analysis

PHA and FMEA are described in Chap. 10 and FTA in Chap. 3. The remaining methods are presented below, separately [3].

11.4.1 Hazard and Operability Studies (HAZOP)

This is a systematic and structured qualitative method that had its beginnings in the chemical process industry [3,9–11]. HAZOP has proven to be an effective approach to identifying unforeseen hazards designed into facilities for various reasons, or hazards introduced into already existing facilities due to factors such as changes made to process conditions or operating procedures.

Three basic objectives of HAZOP are as follows [11]:

- Produce a complete facility/process description.
- Review all facility/process elements to determine how deviations from the design intentions can occur.
- Decide whether the above deviations can lead to operating hazards/problems.

A HAZOP study can be performed by following the five steps listed below [11, 12].

- **Establish study objectives and scope.** This is concerned with developing study objectives and scope by considering all relevant factors.

- **Form HAZOP Team.** This involves forming a HAZOP team composed of persons from design and operation with appropriate experience to determine all possible effects of deviations from intended applications.
- **Collect all types of relevant information.** This involves obtaining the necessary documentation, drawings, and process description. More specifically, it includes items such as equipment specifications, layout drawings, process control logic diagrams, operating and maintenance procedures, and emergency response procedures.
- **Analyze all major pieces of equipment and supporting items.** This involves performing analysis of all major items of equipment and all supporting equipment, piping, and instrumentation using the documents obtained in the previous step.
- **Document the study.** This is the final step and is concerned with documenting items such as consequences of any deviation from the norm, a summary of deviations from the norm, and a summary of deviations considered credible and hazardous.

Three main advantages of HAZOP are as follows [3]:

- Very good track record of previous use and success.
- Does not require specialized tools or extensive training.
- Can produce detailed and comprehensive results.

In contrast, two main drawbacks of HAZOP are that it can be quite time consuming for large and complex systems and it is best for short time periods only because team members can loose effectiveness. The application of HAZOP to a mining system is demonstrated in Ref. [3].

11.4.2 Event Tree Analysis

This is a "bottom up" approach to identifying the possible outcomes when the probability of occurrence of the initiating event is known. The method has proven to be an effective tool for analyzing facilities having engineered accident-mitigating characteristics for identifying the sequence of events that follow the initiating event and produce given sequences. Normally, it is assumed that each sequence event is either a failure or a success.

Because of the inductive nature of this approach, the fundamental question addressed is "What happens if . . . ?" Usually, ETA is used to perform analysis of more complex systems than those handled by the failure mode and effect analysis approach.

Two main advantages of ETA are that it is an effective tool for single events with multiple outcomes and for high risks not amenable to simpler analysis methods [3]. In contrast, its disadvantages include an extremely time-consuming approach and that the event occurrence probabilities may be difficult to estimate. Additional information on ETA is available in Refs. [11–13].

11.4.3 Interface Analysis

Interface analysis is used to determine the incompatibilities between subsystems and assemblies of an item/product that could result in serious accidents. The analysis establishes that distinct units/parts can be integrated into a fairly viable system and the normal operation of a part/unit will not impair the performance or damage another part/unit or the entire system.

The relationships considered by the analysis can be grouped under three areas, shown in Fig. 11.3 [11, 14].

Flow relationships are concerned with two or more units where flow between these units may involve air, steam, water, electrical energy, fuel, steam, or lubricant oil. In addition, flow could be unconfined, such as heat radiation from one item to another. The frequent flow-related problems associated with many products are the proper level of flow of fluids and energy from one unit to another through confined passages. In turn, these may result, directly or indirectly, in safety problems.

Functional relationships are concerned with multiple items. For example, in a situation where outputs of an individual item constitute the inputs to a downstream item, any deficiency in inputs and outputs may cause damage to the downstream item and, in turn, a safety hazard. The outputs could be in conditions such as unprogrammed, zero, excessive, degraded, and erratic outputs.

Physical relationships are concerned with the physical aspects of items. For example, two units/items may be well designed and manufactured and operate quite effectively individually, but they may fail to fit together due to dimensional differences or they may present other difficulties that may lead to safety-related problems.

Some of the advantages of the interference analysis are applicable to all types of systems and interfaces as well as at the subsystem to the component level. In contrast, some of its drawbacks are that it is difficult to apply to complex systems and it is difficult to discover all interface-related incompatibilities for each and every operation.

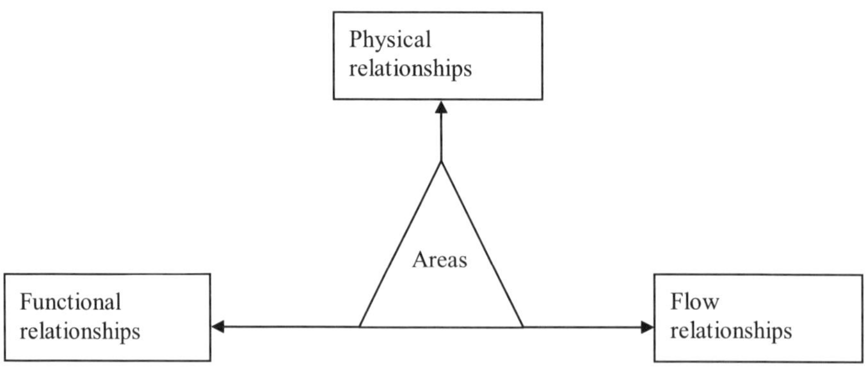

Fig. 11.3 Main areas of relationships considered in interface analysis

11.4.4 Action Error Analysis (AEA)

This is a quite effective method used to identify operator errors and their consequences. The method focuses on the interactions between humans and a system during testing, operation, and maintenance phases. For operation and maintenance tasks AEA is performed by following the three basic steps shown in Fig. 11.4 [15].

In step 3, for each action the following types of errors are considered:

- Incorrect sequence of actions
- Wrong actions taken
- Error of omission (*i.e.*, failure to an action)
- Temporal errors (*i.e.*, actions taken late or early)
- Actions applied to the incorrect object

The main advantage of AEA is that it is well suited for automated or semiautomated processes with operator interfaces; the main disadvantage is that it requires a high level of expertise concerning the system in question.

11.4.5 Operating and Support Analysis (OASA)

This method seeks to identify potential hazards during operation and maintenance, find the associated root causes, determine the acceptable risk level, and recommend appropriate measures for risk reduction. The operating and support analysis is performed by a group of individuals familiar with the system's operation and interaction with all involved humans.

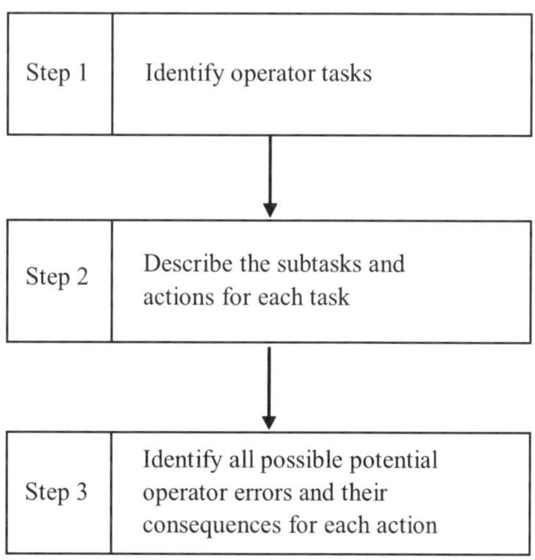

Fig. 11.4 Steps for performing action error analysis for operation and maintenance tasks

Some of the factors considered during OASA are as follows [3, 15]:

- Operation in normal and abnormal situations
- Testing of systems
- Making the required changes to the system
- Training operation and maintenance personnel
- Maintaining the equipment and its associated software
- Providing appropriate documentation for the systems under consideration.

The main advantage of OASA is that it provides hazard identification in the context of the entire system operation. Its major disadvantage is that it requires a high degree of expertise concerning the system in question.

11.4.6 Sequentially Timed Events Plot (STEP) Investigation System

STEP is a useful analytical method that graphically shows sequentially timed events. All events are defined with formatted "building blocks" made up of an "actor and action" [3]. STEP analysis is quite helpful for discovering and analyzing problems and assessing mitigation options. It is also used to perform analysis of the types and sequences of events that can result in an incident.

The two main advantages of STEP are that it can be used to define and systematically analyze complex processes or systems and that it facilitates focus-group analysis. In contrast, its main disadvantage is that it is usually perceived as a complicated approach that is costly to implement.

11.4.7 Potential or Predictive Human Error Analysis

This is a team-based method and is quite similar to the hazard and operability studies (HAZOP) approach. However, this method focuses on human tasks and their associated error potential [16]. The method classifies the causes of human error into five basic categories (along with their corresponding effects in parentheses): complexity (increases the likelihood of error), environment (adverse environments increase the likelihood of error), stress (increases the likelihood of error), fatigue (increases the likelihood of error), and training (better training decreases the likelihood of error).

Therefore, it is important that the members of the team performing the analysis consider these causes of error during the analysis. The basic analysis procedure is composed of the following two steps [3]:

- **Step 1:** Identify key human tasks.
- **Step 2:** For each task, apply guide words such as incomplete action, incorrect action, incorrect selection, action omitted, wrong action sequence, wrong action timing, and action applied to incorrect interface object.

The main advantage of the method is that it can identify a high proportion of potential errors. Similarly, its two main disadvantages are that it can be a quite time-consuming approach if there are many tasks and actions and its effectiveness depends on the team's expertise and effort [3].

11.5 Lessons Learned in Addressing the Safety of Programmable Electronic Mining Systems

Over the years, the National Institute for Occupational Safety and Health (NIOSH), in conjunction with Mine Safety and Health Administration (MSHA) and the industrial sector, generated the NIOSH safety framework of functional safety of PE mining systems. In this regard, many valuable lessons were learned for addressing functional safety and for changing the perspectives and practices of the industrial sector [1]. These lessons are benefiting the mining industry and can also be applied to other industries as well. Most of these lessons are as follows [1]:

- Involve the industrial sector early and on a continuous basis.
- Identify and understand all related issues and perceptions.
- Establish important concepts, terminology, and definitions as early as possible.
- Decompose the problem under consideration into manageable parts.
- Separate the associated concerns.
- Conduct industry workshops as considered appropriate.
- Make use of scenarios for conveying some types of information.

11.6 Obtaining Programmable Electronic Mining System Safety-Related Information

There are many sources for obtaining, directly or indirectly, programmable electronic mining system safety-related information. This section presents a number of such sources classified under three categories.

11.6.1 Organizations and Systems

- National Institute for Occupational Safety and Health (NIOSH), 200 Independence Avene SW, Washington, DC, USA
- Occupational Safety and Health Administration (OSHA), US Department of Labor, 200 Constitution Avenue, Washington, DC, USA
- The American Society of Safety Engineers, 1800 E. Oakton Street, Des Plaines, IL, USA

- System Safety Society, 14252 Culver Drive, Suite A-261, Irvine, CA, USA
- Institute of Electrical and Electronics Engineers (IEEE), 445 Hoes Lane, Piscataway, NJ 08854-4141, USA
- International Electro-technical Commission (IEC), 3 rue de Varembe, P.O. Box 131, CH-11, Geneva, Switzerland
- National Safety Council, 444 North Michigan Avenue, Chicago, IL, USA
- National Electronic Injury Surveillance System, US Consumer Product Safety Commission, 5401 Westbard Street, Washington, DC, USA
- Computer Accident/Incident Report System, System Safety Development Center, EG8G, PO Box 1625, Idaho Falls, ID, USA
- GIDEP Data, Government Industry Data Exchange Program (GIDEP) Operations Center, Fleet Missile Systems, Analysis, and Evaluation, Department of Navy, Corona, CA, USA

11.6.2 Books and Standards

- Levenson, N.G.: Safeware: System Safety and Computers, Addison-Wesley, Reading, MA (1995)
- Perrow, C.: Normal Accidents: Living with High-Risk Technologies, Princeton University Press, Princeton, NJ (1999)
- Beizer, B.: Software Testing Techniques, International Thomson Computer Press, London (1990)
- Redmill, F., Chudleigh, M., Catmur, J.: System Safety: HAZOP and Software HAZOP. Wiley, New York (1999)
- IEC 61508 SET, Functional Safety of Electrical/Electronic/Programmable Electronic Safety-Related Systems, Parts 1–7, International Electrotechnical Commission, Geneva, Switzerland (2000)
- Defence Standard 00-58, Parts I and II, HAZOP Studies on Systems Containing Programmable Electronics, Directorate of Standardization, UK Ministry of Defence, Glasgow, UK (1998)
- MIL-STD-882C, System Safety Program Requirements, US Department of Defense, Washington, DC (1993)
- IEEE-STD-1228, Standard for Software Safety Plans, Institute of Electrical and Electronics Engineers (IEEE), Piscataway, NJ (1994)
- IEEE-STD-730, Standard for Software Quality Assurance Plans, Institute of Electrical and Electronics Engineers, Piscataway, NJ (1995)
- IEEE-STD-830, Recommended Practice for Software Requirement Specifications, Institute of Electrical and Electronics Engineers (IEEE), Piscataway, NJ (1993)

11.6.3 Commercial Sources for Obtaining Standards

- Global Engineering Documents,
  15 Inverness Way East,
  Englewood, CO 80112-5704
  USA
- Total Information Inc.,
  844 Dewey Avenue,
  Rochester, NY 14613
  USA
- Document Center, Inc.,
  111 Industrial Road, Suite 9,
  Belmont, CA 94002
  USA

11.7 Problems

1. What is a programmable electronic mining system?
2. What are the main areas of programmable electronics usage in mining?
3. What are the major factors contributing to programmable-electronics-based mishaps in the area of mining?
4. List at least six methods for performing hazard and risk analysis of programmable electronic mining systems.
5. Describe the following two methods:
   - Operating and support analysis
   - Action error analysis
6. Discuss the lessons learned in addressing the safety of programmable electronic mining systems.
7. List the five most important sources for obtaining programmable electronic mining system safety-related information.
8. Compare interface analysis with event tree analysis.
9. Discuss the advantages and disadvantages of hazard and operability studies (HAZOP).
10. Discuss potential or predictive human error analysis.

References

1. Sammarco, J.J.: Addressing the safety of programmable electronic mining systems: lessons learned. In: Proceedings of the 37th IEEE Industry Applications Society Meeting, pp. 692–698 (2003)

2. Sammarco, J.J., Kohler, J.L., Novak, T., Morley, L.A.: Safety issues and the use of software-controlled equipment in the mining industry. In: Proceedings of the 32nd IEEE Industry Applications Society Meeting, pp. 496–502 (1997)
3. Sammarco, J.J.: Programmable electronic mining systems: best practice recommendations (in nine parts), Report No. IC 9480 (Part 6: 5.1 System Safety Guidance), National Institute for Occupational Safety and Health (NIOSH), US Department of Health and Human Services, Washington, DC (2005). Available from NIOSH: Publications Dissemination, 4676 Columbia Parkway, Cincinnati, OH 45226, USA (1998)
4. Sammarco, J.J.: Safety framework for programmable electronics in mining. Min. Eng. 51, 12, 30–33 (1999)
5. Dransite, G.D.: Ghosting of electro-hydraulic long wall shield advanced systems. In: Proceedings of the 11th West Virginia University International Electrotechnology Conference, pp. 77–78 (1992)
6. Fatal Alert Bulletins, Fatal Grams and Fatal Investigation Reports, Mine Safety and Health Administration (MSHA), Washington, DC (May 2001). Available online at www.msha.gov/fatals/fab.htm
7. Waudby, J.F.: Underground coal mining remote control of mining equipment: known incidents of unplanned operation in New South Wales (NSW) underground coal mines. Dept. of Mineral Resources, NSW Department of Primary Industries, Maitland, NSW, Australia (2001)
8. IEC61508, Parts 1–7, Functional safety of electrical/electronic/ programmable electronic safety-related systems. International Electrotechnical Commission, Geneva (1998)
9. Defense Standard 00-58 (Parts I and II), HAZOP Studies on Systems Containing Programmable Electronics, Directorate of Standardization, UK Ministry of Defence, Glasgow, UK (1998)
10. Redmill, F., Chudleigh, M., Catmur, J.: System Safety: HAZOP and Software HAZOP. Wiley, New York (1999)
11. Dhillon, B.S.: Engineering Safety: Fundamentals, Techniques, and Applications. World Scientific, River Edge, NJ (2003)
12. CAN/CSA-Q6340-91, Risk Analysis Requirements and Guidelines, prepared by the Canadian Standards Association, Toronto (1991). Available from the Canadian Standards Association (CSA), 178 Rexdale Boulevard, Rexdale, Ontario, Canada
13. Cox, S.J., Tait, N.R.S.: Reliability, Safety, and Risk Management. Butterworth-Heinemann, Oxford (1991)
14. Hammer, W.: Product Safety Management and Engineering. Prentice-Hall, Englewood Cliffs, NJ (1980)
15. Harm-Ringdahl, L.: Safety Analysis: Principle and Practice. Elsevier, London (1993)
16. Guidelines for Preventing Human Error in Process Safety, Center for Chemical Process Safety, American Institute of Chemical Engineers, New York (1994)

Bibliography: Literature on Mining Equipment Reliability, Maintainability, and Safety

Introduction

Over the years, a large number of publications relating to mining equipment reliability, maintainability, and safety have appeared in the form of journal articles, conference proceedings articles, technical reports, etc. This appendix presents an extensive list of such publications.

The period covered by the listing is 1965–2007. The main objective of this listing is to provide readers with sources for obtaining additional information on mining equipment reliability, maintainability, and safety.

Publications

1. Akashev, Z.T., Atagulov, B.E.: Organizational indexes of quality of repair of quarry locomotives, *Tyazheloe Mashinostroenie* 3, 25–27 (2005)
2. Aldinger, J.A., Keran, C.M.: A review of accidents during surface mine mobile equipment operations. In: *Proc. of the 25th Annual Institute on Mining Health, Safety, and Research*, pp. 99–108 (1994)
3. Altshuler, V.M.: A method of constructing a mathematical model to study the reliability of mine transportation systems. *Sov. Min. Sci.* 5, 72–76 (1969)
4. Ambs, J.L., Setren, R. S.: Safety evaluation of disposable diesel exhaust filters for permissible mining equipment. In: *Proceedings of the 7th US Mine Ventilation Symposium*, pp. 105–110 (1995)
5. Anderson, W.D., Silbert, M.N., Lloyd, J.R.: Reliability procedure for fixed offshore platforms. *J. Struct. Div. Am. Soc. Civil Eng.* 108, STll, 2517–2538 (1982)
6. Angelides, D.C., Vereziano, D., Sunder, S.S.: Random sea and reliability of offshore foundations. *J. Eng. Mechan. Div. Am. Soc. Civil Eng.* 107, EMl, 131–148 (1981)

7. Anon.: Environmental safety underground. *Colliery Guardian* 235, 5, 170–172 (1987)
8. Anon.: High altitude reliability for heavy mining vehicles. *Austral. Min.* 85, 4, 20–22 (1993)
9. Anon.: Recent mine safety R&D projects of the U.S. Bureau of Mines. *Eng. Min. J.* 196, 11, 26–28 (1995)
10. Anon.: U.S. enacts new mine safety policies. *Coal Age* 111, 6, 24–27 (2006)
11. Anon.: Conveying safety, *Min. Mag.* 193, 4, 24–26 (2005)
12. Anon.: Crane Hire Group helps LIFT mine maintenance rates. *Austral. Min.* 98, 4, 38 (2006)
13. Anon.: Emerson's machinery health technologies enable UK coal to avoid equipment failures. *Coal Int.* 254, 3, 42–43 (2006)
14. Anon.: Improvements to cat machine safety. *Min. Mag.* January, 45 (2006)
15. Anon.: Management in action-new versus used equipment: know your supplier. *Min. Mag.* 194, 3, 28–29 (2006)
16. Anon.: Management in action-safety: still unacceptable. *Min. Mag.* 194, 4, 22–26 (2006)
17. Anon.: Minimizing maintenance with oil condition monitoring. *Coal Age*, 110, 9, 36–37 (2005)
18. Anon.: Monitoring mining equipment. *Eng. Australia* 77, 4, 59–60 (2005)
19. Anon.: New products aim at cutting equipment maintenance costs. *Eng. Min. J.* 207, 3, 60–63 (2006)
20. Anon.: Stop bolting hazards in their tracks. *Austral. Min.* 98, 6, 34 (2006)
21. Anon.: The color of safety. *Eng. Min. J.* 207, 5, 46–49 (2006)
22. Astafyev, Y.: Simulation ore reserves control model based on reliability characteristics of open-pit equipment. In: *Proc. of the Symposium on Applications of Computer Methods in the Mining Industry*, pp. 10–14 (1976)
23. Bakewell, D.: The reliability problem. *Min. Technol.* 64, 736, 50–53 (1982)
24. Barabady, J.: Reliability and maintainability analysis of crushing plants in Jajarm bauxite mine of Iran. In: *Proc. of the Annual Reliability and Maintainability Symposium*, pp. 109–115 (2005)
25. Barczak, T., Gearhart, D.: Performance and safety considerations of hydraulic support systems. In: *Proc. of the 17th Int. Conf. on Ground Control in Mining*, pp. 176–186 (1998)
26. Barkand, T.D.: Emergency braking systems for mine elevators. In: *Proc. of the Conf. on New Technology in Mine Health and Safety*, pp. 325–336 (1992)
27. Barlow, P.: The economics, reliability and versatility of high-angle conveyors. *Quarry Manage.* 21, 4, 17–23 (1994)
28. Barton, J.L., Murray, R.A.: Application of conveyor belt monitoring technology to conveyor belt reliability. *CIM Bull.* 85, 959, 37–44 (1992)
29. Bates J.J.: The technical performance and reliability of powered support installations. *Min. Technol.* 59, 679, 168–175 (1977)
30. Bates, J.J.: Technical commissioning of coal face equipment. *Min. Technol.* 55, 628, 52–62 (1973)

31. Bates, J.J.: The key technical factors of machine performance and reliability – No. 1. *Min. Technol.* 63, 725, 31–33 (1981)
32. Bates, J.J.: The key technical factors of machine performance and reliability – No. 2. *Min. Technol.* 63, 724, 31–33 (1981)
33. Bates, J.J.: The key technical factors of machine performance and reliability – No. 3. *Min. Technol.* 63, 725, 105–107 (1981)
34. Bates, J.J.: The key technical factors of machine performance and reliability – No. 4. *Min. Technol.* 63, 726, 159–160 (1981)
35. Bates, J.J.: The key technical factors of machine performance and reliability – No. 5. *Min. Technol.* 64, 742, 383–385 (1982)
36. Bates, J.J.: The key technical factors of machine performance and reliability (second series: Part I). *Min. Technol.* 66, 759, 23–25 (1984)
37. Bates, J.J.: The key technical factors of machine performance and reliability (second series: Part II). *Min. Technol.* 66, 760, 67–60 (1984)
38. Bates, J.J.: The key technical factors of machine performance and reliability (second series: Part III). *Min. Technol.* 66, 761, 99–104 (1984)
39. Bates, J.J.: The key technical factors of machine performance and reliability (second series: Part IV). *Min. Technol.* 66, 762, 147–152 (1984)
40. Bates, J.J.: The key technical factors of machine performance and reliability (second series: Part V). *Min. Technol.* 66, 766, 277–281 (1984)
41. Bea, R.G.: Reliability considerations in offshore platform criteria. *J. Struct. Div. Am. Soc. Civil Eng.* 106, ST9, 1835–1853 (1980)
42. Bea, R.G.: Selection of environmental criteria for offshore platform design. *J. Petroleum Technol.* 1206–1214 (1974)
43. Bello, G.C., Avogadsi S.: Offshore pipeline system – how to decide the optimal redundancies using reliability engineering. *Reliabil. Eng.* 3, 89–99 (1982)
44. Benger, H.: Mine ventilation system with automatically exchangeable active parts. *Brown Boveri Rev.* 73, 4, 204–210 (1986)
45. Beus, M.J., Iverson, S.: Safer mine hoisting with conveyance position and load monitoring. *Am. J. Ind. Med.* 36, 119–121 (1999)
46. Bickel, K.L.: Five Automatic Fire Protection Systems for Mobile Mining Equipment: Their Design and in Mine Reliability Testing, US Bureau of Mines, Washington, DC, pp. 14–17 (1977)
47. Bleazard, D., Hepler, D., Dearman, L.: Equipment reliability improved at Barrick goldstrike. *Min. Eng.* 50, 1, 45–49 (1998)
48. Bobick, T.G.: Increasing safety in underground coal mining through improved materials-handling systems. In: *Proc. of the SME Annual Meeting*, pp. 1–9 (1988)
49. Bondar, S.A., Meruov, V.V.: Operational reliability of conveyer lines with intermediate storage capacity. *Sov. Min. Sci.* 15, 3, 268–270 (1979)
50. Bozorgebrahimi, A., Hall, R.A., Morin, M.A.: Equipment size effects on open pit mining performance. *Int. J. Surface Min. Reclamat. Environ.* 19, 1, 41–56 (2005)

51. Brahma, K.C., Pal, B.K., Das, C.: A study on application of markovian decision process in drilling and blasting operation of opencast mines. *J. Mines Metals Fuels* 55, 1–2, 24–29 (2007)
52. Brodzinski, S.: Reliability of hydraulic disc brakes in winders. *Colliery Guardian* February, 67–69 (1984)
53. Burgess, T., Shimbel, L.: What is the prognosis maintenance program? *Eng. Min. J.* 196, 5, 32–35 (1995)
54. Butani, S.J.: Hazard analysis of mining equipment by mine type and geographical region. In: *Proc. of the Symp. on Engineering Health and Safety*, pp. 158–173 (1986)
55. Butosin, Yu.M., Gavrilenko, V.A.: Some aspects of the reliability of drilling euipment. *Sov. Min. Sci.* 12, 62–65 (1976)
56. Castanier, B., Rausand, M.: Maintenance optimization for subsea oil pipelines. *Int. J. Pressure Vessels Pip.* 83, 236–243 (2006)
57. Cawley, J.C.: Electrical accidents in the mining industry, 1990–1999. In: *Proceedings of the 36th IAS Annual Meeting*, 2, 1361–1368 (2001)
58. Cawley, J. C.: Probability of Spark Ignition in Intrinsically Safe Circuits, Report No. 9183, *Report of Investigations*, US Bureau of Mines, Washington DC, pp. 1–15 (1988)
59. Clark, D.: Tribology – its application to equipment reliability and maintainability design in the underground coal mining industry. In: *Proc. of the Int. Tribology Conference*, pp. 38–44 (1990)
60. Clark, R., Dawkins, G., Kennedy, R., Grainger, A., Lysaght, A.: Maintenance performance improvement – a structured approach. In: *Proc. of the Int. Conf. on Reliability, Production, and Control in Coal Mines*, pp. 89–94 (1991)
61. Collins, J., Singhal, R.K., Fytas, K.: Mine maintenance analysis through Markov chain schemes. *Int. J. Surface Min. Reclamat.* 6, 1, 47–56 (1992)
62. Committee on Reliability of Offshore Structures, Application of Reliability Methods in Design and Analysis of Offshore Platforms, *J. Struct. Eng.* 109, 10, 2265–2291 (1983)
63. Conway, E.J., Unger, R.L.: Maintainability design of mobile underground mining equipment. In: *Proc. of the Bureau of Mines Technology Transfer Seminar*, pp. 61–65 (1987)
64. Conyngham, M.G.: Role of quality assurance in manufacturing of mining equipment. In: *Proc. of the Int. Conf. on Reliability, Production, and Control in Coal Mines*, pp. 95–103 (1991)
65. Cooper, B.: Maintenance strategy procedures development and implementation. *Min. Technol.* 78, 893, 3–6 (1996)
66. Cox, D.: The maintenance and repair of explosion-protected equipment in coal mines. *Vector (Electr. Eng.)* October, 19–22 (1998)
67. Cutts, A.: Improving the reliability of machines and systems used in the production processes, automation and control technologies. In: *Proc. of the International Conference on Reliability, Production, and Control in Coal Mines*, pp. 18–24 (1991)

68. Czaplicki, J.M., Brodzinski, S.: Reliability Testing of Polish Winders. *Colliery Guard. Coal Int.* January, 38–39 (1980)
69. Daniel, J.H.: Diesels – Backbone of a changing mining industry. In: *Proc. of the 4rth US Mine Ventilation Symposium*, pp. 561–568 (1989)
70. Daniel, J.H.: Reducing mine accidents by design. In: *Proc. of the SME Annual Meeting*, pp. 1–11 (1991)
71. Dasgupta, K., Chattapadhyay, A., Mondal, S.K.: Selection of fire resistant hydraulic fluids through system modeling and simulation. *Simulat. Modell. Pract. Theory* 13, 1, 1–20 (2005)
72. Davies, R.K.L.: Monitoring hoist safety equipment. *Coal Age* 21, 1, 66–68 (1986)
73. Dawson, V.E., Heistad, R.: Performance and field testing of the service brakes of mining trucks in B.C. *Trans. Can. Inst. Min. Metallurgy* LXXX, 129–136 (1977)
74. De La Mare, R.F.: Failure characteristics of mechanical equipment which receives comprehensive maintenance overhaul. *Qual. Reliabil. Eng. Int.* 5, 3, 189–202 (1989)
75. De Rosa, M.I.: Equipment fires cause injuries. *Coal Age* 109, 10, 28–31 (2004)
76. DeGaspari, J.: Armchair mining. *Mech. Eng.* 125, 5, 42 (2003)
77. Dey, A., Bhattacharya, J., Banerjee, S.: Prediction of field reliability for dumper tyres. *Int. J. Surface Min. Reclamat.* 8, 1, 23–25 (1994)
78. Dhillon, B.S.: Bibliography of literature on mining equipment reliability. *Microelectron. Reliabil.* 26, 6, 1131–1138 (1986)
79. Dittrich, W., Dralle, H.: Inspection of opencast mine equipment. *World Min.-Surface Underground* 59, 1, 22–26 (2007)
80. Dkilov, K.N.: Reliability of the technology of extracting seams with complex geology. *Sov. Min. Sci.* 16, 3, 242–247 (1980)
81. Dotson, J.C.: Reliability engineering and its application in mining. *Min. Congr. J.* 52, 8, 68–76 (1966)
82. Dreyer, E.: Cost-effective prevention of equipment failure in the mining industry. *Int. J. Pressure Vessels Pip.* 61, 2–3, 329–347 (1995)
83. Dunn, S.: Optimizing production scheduling for maximum plant utilization and minimum downtime, the reliability revolution. In: *Proc. of the Dollar Driven Mining Conference*, Perth, Western Australia, pp. 8–12 (July 1997)
84. Dunzgun, H.S.B., Einstein, H.H.: Assessment and management of roof fall risks in underground coal mines. *Safety Sci.* 42, 1, 23–41 (2004)
85. Dvornikov, L.T., Turov, V.A.: Composition and structure of reliability properties. *Sov. Min. Sci.* 24, 5, 448–451 (1989)
86. Dvornikov, L.T., Gudimov, N.I., Turov, V., Davydov, A.P.: Reliability analysis of BGA rotary – percussive units for drilling boreholes in rock. *Sov. Min. Sci.* 15, 4, 365–370 (1979)
87. Earthrowl, B.: Design for reduced maintenance costs – the application of split roller bearings in mining equipment. *Min. Technol.* 76, 876, 168–170 (1994)

88. Ercelebi, S.G., Yegulalp, T.M.: Reliability and availability analysis of mining systems. *Trans. Inst. Min. Metallurgy* 102, 51–58 (1993)
89. Erskine, W.W.: Qualities necessary for engineering management. *Min. Technol.* September, 392–395 (1982)
90. Esterhuizen, S.G., Gurtunca, R.G.: Coal mine safety achievements in the USA and the contribution of NIOSH research. *J. South African Inst. Min. Metallurgy* 106, 12, 813–820 (2006)
91. Feder, D.K.: Preventive maintenance of open-cut continuous mining equipment. *Weld. World* 32, 1, 267–276 (1993)
92. Feltes, M.J.: Mechanical belt fasteners increase conveyor availability. *Min. Eng.* 45, 3, 273–274 (1993)
93. Forsman,B., Kumar, U.: Surface mining equipment and maintenance trends in the Scandinavian countries. *J. Mines Metals Fuels* 40, (8–9), 266–269 (1992)
94. Freeman, S.: Remote blasting lifts safety and output. *Austral. Min.* 97, 7, 38–39 (2005)
95. Freidina, E.V., Kovalenko, A.S., Rudenko, O.A.: Effect of mine-transport-equipment reliability on the productivity of a quarry system. *Sov. Min. Sci.* 10, 50–54 (1975)
96. Gel'Fand, F.M., Bur'ka, V.F., Shul'gin, E.I.: Probabilistic method of finding burst hazard factor or index and its limiting values for safe and hazardous seams. *Sov. Min. Sci.* 15, 3, 275–278 (1979)
97. Gibiec, M.: Prediction of machine health with application of an intelligent approach: a mining machinery case study. *Key Eng. Mater.* 293/294, 661–668 (2005)
98. Gimel'shein, L.Y., Nikeshin, B.S.: Quantitative reliability criteria for electromechanical systems in mines. *Sov. Min. Sci.* 3, 199–202 (1967)
99. Girard, J., McHugh, E.: Detecting Problems with Mine Slope Stability. In: *Proc. of the 32nd Annual Institute on Mining Health, Safety, and Research*, pp. 1–8 (2000)
100. Glavin, J.: Health and safety in Australia's mines. *Mater. World*, 14, 4, 22–23 (2006)
101. Goldheck, L.: Stored energy in conveyor belts: managing the risk. *Eng. Min. J.* 207, 10, 56–59 (2006)
102. Gordon, R.P.E., Flin, R.H., Mearns, K., Fleming, M.T.: Assessing the Human Factors causes of accidents in the offshore oil industry. In: *Proc. of the Int. Conf. on Health, Safety, and Environment in Oil Gas Exploration and Production*, pp. 635–644 (1996)
103. Gorzalczynski, S.: Limitations of safety arrest mechanisms for mine shaft conveyances. *CIM Bull.* 95, 1065, 67–71 (2002)
104. Gupta, S., Ramkrishna, N., Battacharya, J.: Replacement and maintenance analysis of longwall shearer using fault tree technique. *Trans. Inst. Min. Metallurgy Sect. A Min. Technol.* 115, 2, 49–58 (2006)
105. Gwinnett, T.: Underground breakdown. *Colliery Guard.* July, 265–266 (1981)

106. Hall, L.: System design and equipment reliability for Wide Web Working at Hem Heath Colliery, *Min. Technol.* 64, 736, 54–57 (1982)

107. Hall, R.A., Daneshmend, L.K.: Reliability and maintainability models for mobile underground haulage equipment. *CIM Bull.* 96, 1072, 159–165 (2003)

108. Hall, R.A., Daneshmend, L.K.: Reliability modelling of surface mining equipment: data gathering and analysis methodologies. *Int. J. Surf. Min. Reclamat. Environ.* 17, 3, 139–155 (2003)

109. Hall, R.A., Daneshmend, L.K., Lipsett, M.G., Wong, J.: Reliability analysis as a tool for surface mining equipment evaluation and selection. *CIM Bull.* 93, 1044, 78–82 (2000)

110. Hartley, D., Jarvis., E.E.: Time or condition-based maintenance. *Min. Eng.* 146, 305, 506–508 (1987)

111. Hartman, H.L., Novak, T., Gregg, A.J.: Health hazards of diesel and electric vehicles in an underground coal mine. *Min. Sci. Technol.* 5, 2, 131–151 (1987)

112. Haskayne, J.D., Farmer, S.D.: The reliability problem – powered supports. *Min. Technol.* 64, 736, 84–87 (1982)

113. Heyns, F., Van Der Westhuizen, J.: A mining case study: the safe maintenance of underground railway track. *Civil Eng.* 14, 5, 8–10 (2006)

114. Heyns, F.J.: Construction and maintenance of underground railway tracks to safety standard of SANS: 0339. *J. So. Afr. Inst. Min. Metallurgy* 106, 12, 793–798 (2006)

115. Hosseini, M.: Reliability revolution. *Can. Min. J.* 120, 2, 33–35 (1999)

116. Howey, R.A., Gaarder, O.H.: Application of reliability engineering to offshore production equipment. *J. Petroleum Technol.* April, 339–346 (1977)

117. Iannacchione, A.T., *et al.*: Controlling roof beam failures from high horizontal stresses in underground stone mines. In: *Proc. of the 17th Int. Conf. on Ground Control in Mining*, pp. 102–112 (1998)

118. Iannacchione, A.T., *et al.*: Safer mine layouts for underground stone mines subjected to excessive levels of horizontal stress. *Min. Eng.* 55, April, 25–31 (2003)

119. Iannaccione, A.T., Prosser, L.J., Esterhuizen, G., Bajpayee, T.S.: Technique to Assess Hazards in Underground Stone Mines: The Roof-Fall-Risk Index (RFRI), *Min. Eng.* 59, 1, 49–57 (2007)

120. Iannachione, A.T., Marshall, T.E., Prosser, L.J.: Failure characteristic of roof falls at an underground stone mine in southwestern Pennsylvania. In: *Proc. of the 20th Int. Conf. on Ground Control in Mining*, pp. 119–125 (2001)

121. Iverson, S., Kerkering, J.C., Coleman, P.: Using fault tree analysis to focus mine safety research. In: *Proc. of the 108th Annual Meeting of the Society for Mining, Metallurgy, and Exploration*, pp. 1–10 (2001)

122. Iyko, V.I., Ovchinnikova, L.K., Plontnikova, V.I.: A method of estimating the operational reliability of kinematic mechanized support systems. *Sov. Min. Sci.* 9, 333–335 (1974)

123. Jaeger, F.T.: Wear – is your maintenance dollar going? In: *Proceedings of the 4th Canadian Institute of Mining Mechanical/Electrical Plant Engineering and Maintenance Conference*, pp. 38–44 (1984)
124. Jain, V.K.: Developments in surface mining technology-planning, equipment maintenance and environmental management. *J. Mines Metals Fuels* 52, 11, 293–294 (2006)
125. Jardine, A.K.S., Banjevic, D., Wiseman, M., Buck, S., Joseph, T.: Optimising a mine haul truck wheel motors' condition monitoring program: use of proportional hazards modelling. *J. Qual. Mainten. Eng.* 7, 4, 286–301 (2001)
126. Jolly, S.M., Deakin, M.P.: Predictive maintenance, a practical approach. *Min. Technol.* 75, 864, 99–104 (1993)
127. Joy, J.: Use of qualitative and quantitative risk analysis techniques to control safety and other losses in underground coal mining. In: *Proc. of the Int. Conf. on Reliability, Production, and Control in Coal Mines*, pp. 333–335 (1991)
128. Kamenskii, M.K., Nikitina, M.A., Muratov, A.S.: Investigation of the operational reliability of grade EVT-1140V mining cables. *Sov. Electr. Eng.* 51, 12, 113–117 (1980)
129. Keran, C.M., Hendricks, P.A.: Automation and safety of mobile mining equipment. *Eng. Min. J.* 196, 2, 30–33 (1995)
130. Kielblock, A.J.: Ergonomics: A Nice-to-Have for the South African Mining Industry or Potential Time-Bomb? *J. Mine Ventilat.* 52, 3, 116–121 (1999)
131. Kincaid, R.L.: Modern mine maintenance: toolbox, blackbox, or tribo-technology? *Int. J. Surf. Min. Reclamat. Environ.* 10, 2, 47–54 (1996)
132. Kohler, J.L., Sottile, J., Trutt, F.C.: Condition-based maintenance of electrical machines. In: *Proc. of the IAS Annual Meeting*, 1, 205–211 (1999)
133. Koketayev, A.I.: Problems of safety when catastrophic failure arises at mines. In: *Proc. of the 12th Int. Symp. on Mine Planning and Equipment Selection*, p. 505 (2003)
134. Kolomiitsov, M.D, Gabov, V.V.: Dependence of the operational reliability of a cutter-loader on the magnitude and character of the load. *Sov. Min. Sci.* 13, 2, 164–167 (1977)
135. Kuklin, L.G., Boyarskikh, G.A., Semakina, T.P.: Assessing the Technological reliability of drilling equipment. *Min. J.* 5, 70–72 (1983)
136. Kulakov, V.N., Frolov, B.A.: Improving the reliability of engineering systems by introducing an effective reserve. *Sov. Min. Sci.* 11, 6, 763–766 (1975)
137. Kumar, N., Paul, B.: Planning of risk assessment and safety management in indian surface mines. *J. Mines Metals Fuels* 52, 11, 314–316 (2006)
138. Kumar, U., Granholm, S.: Some simple graphical methods for reliability analysis in mines. In: *Proc. of the Int. Conf. on Reliability, Production, and Control in Coal Mines*, pp. 36–41 (1991)
139. Kumar, U., Klefsjoe, B.: Reliability analysis of hydraulic systems of LHD machines using the power law process model. *Reliabil. Eng. Syst. Safety* 35, 3, 217–224 (1992)

140. Kumar, U.: Maintenance management for modern mining equipment. *J. Mines Metals Fuels* 44, 1, 25–29 (1996)
141. Kuruppu, M.: Methods and reliability of measuring winder rope degradation. In: *Proc. of the 12th Int. Symp. on Mine Planning and Equipment Selection*, 1, 261–266 (2003)
142. Kuznetsov, A.P.: Attaining statistical reliability in models of random properties of a continuous medium. *Sov. Min. Sci.* 9, 5, 573–575 (1973)
143. Kwitowski, A.J., Brautigam, A.L., Monaghan, W.D.: Teleoperated continuous mining machine for improved safety. *Min. Eng.* 47, 8, 753–759 (1995)
144. Kyriazi, N., Kovac, J., Duerr, W., Shubilla, J.: Laboratory Testing of Compressed – Oxygen Self – Rescuers for Ruggedness and Reliability, US Bureau of Mines, Report of Investigation 8839, Washington, DC (1983)
145. Lauterback, T.: Management in action: safety. *Min. Mag.* 194, 4, 28–29 (2006)
146. Lawrence, A.C.: Human error as a cause of accidents in gold mining. *J. Safety Res.* 6, 2, 78–88 (1974)
147. Leakovich, P.E., Chalenko, N.E.: Use of reliability theory to calculte the required number of reserve longwall faces. *Sov. Min. Sci.* 5, 2, 160–164 (1969)
148. Leugner,T.: Developing a total productive maintenance (TPM) programme. *Quarry Manage.* 23, 8, 21–25 (1996)
149. Lewis, M.W., Steinberg, L.: Maintenance of mobile mine equipment in the information age. *J. Qual. Mainten. Eng.* 7, 4, 264–274 (2001)
150. Liang, X.: Application of casuality diagram in fault analysis of machine equipment in coal mines. *Jisuanji Gongcheng/Comput. Eng.*, 31, 5, 204–209 (2005)
151. Lipowczan, A.: Increasing the reliability and safety of mining machines by application of the vibration diagnostic (experiences and results). In: *Proc. of the Int. Conf. on Reliability, Production, and Control in Coal Mines*, pp. 155–163 (1991)
152. Litton, C.D.: Improving fire detection in underground coal mines. In: *Proc. of the SME Annual Meeting*, pp. 216–226 (1986)
153. Lloyd, M.R.: Engineering testing, assessment and research for equipment safety in New South Wales coal mines. *Min. Technol.* 68, 792, 307–308 (1986)
154. Louit, D.M., Knights, P.F.: Simulation of initiatives to improve mine maintenance. *Trans. Inst. Min. Metallurgy* 110, 47–58 (2001)
155. Lysenko, V.N., Pavlovskaya, S.A.: A method for estimating the reliability of burst-prevention measures. *Sov. Min. Sci.* 14, 3, 319–320 (1978)
156. Ma, Y., Song, H., Sun, B.: Research on the reliability of mining production systems and optimal maintenance policy. In: *Proc. of the 3rd Int. Symp. on Mine Planning and Equipment Selection*, pp. 451–454 (1994)
157. Majumdar, S.K.: Study on reliability modelling of a hydraulic excavator system. *Qual. Reliabil. Eng. Int.* 11, 1, 49–63 (1995)
158. Makhinin, A.I.: The reliability and service life of the MS-50m centrifugal pit pump, *Sov. Min. Sci.* 5, 6, 719–722 (1969)

159. Marquez, A.C.: Modeling critical failures maintenance: a case study for mining. *J. Qual. Mainten. Eng.* 11, 4, 301–317 (2005)

160. Marshall, P.W.: Risk evaluation for offshore structures. *J. Struct. Div. Am. Soc. Civil Eng.* 95, ST12, 2907–2929 (1969)

161. Mason, N.S.: Monitoring the reliability of coalface equipment. *Min. E* 143, 264, 105–112 (1983)

162. May, J.P., Aldinger, J.A.: Overview of bureau research directed towards face powered haulage safety. In: *Proc. of the 26th Annual Institute on Mi Health, Safety, and Research*, pp. 121–131 (1995)

163. McIvor, R.A.: Mine shaft conveyance safety mechanism – free-fall testi *CIM Bull.* 89, 1004, 47–50 (1996)

164. Miller, R.E., Lowe, N.T., Thompson, R.: A GPS based system for mini ing jolts to heavy equipment operators. In: *Proc. of the Society of Autom Engineer (SAE) Commercial Vehicle Engineering Congress and Exhibit*, pp. 1–4 (2004)

165. Miller, S.M., *et al.*: Computer modelling of catch benches to mitigate r fall hazards in open pit mines. In: *Proc. of the 4th North American Mechanics Symposium*, pp. 539–545 (2000)

166. Mohan, S., Duarte, D.: Cognitive modelling of underground miners resp to accidents. In: *Proc. of the Annual Reliability and Maintainability Sy sium*, pp. 51–56 (2006)

167. Moody, C.: Reliable conveyor belt design. In: *Proc. of the American Mi Congress Coal Convention*, 579–582 (1991)

168. Moore, K.: Testing and reliability. *Min. Eng.* 142, 257, 457–462 (1983)

169. Morrell, H.W.: European standards – mining machinery safety. *Min. Technol.* 74, 851, 13–14 (1992)

170. Moses, F., Stahl, B.: Reliability analysis format for offshore structures. *J. Petroleum Technol.* March, 347–354 (1979)

171. Mosinets, V.N.: Assessment of new mining technologies using the theory of reliability. *Sov. Min. Sci.* 1, 4, 335–342 (1965)

172. Mrig, G.C.: Reliability of machines and systems used in production process in mines – a case study of South Eastern Coalfields Limited, India. In: *Proc. of the Int. Conf. on Reliability, Production, and Control in Coal Mines*, pp. 104–110 (1991)

173. Mukherjee, P.S.: Analysis of lubricant use in mining equipment – a tool for maintenance management. *Ind. Lubricat. Tribol.* 53, 6, 256–260 (2001)

174. Mukhopadhyay, A.K.: Open pit system reliability. *J. Mines Metals Fuels* 36, 8, 389–392 (1988)

175. Muller, C., Scharmach, M., Gaal, M., Guelle, A., Lewis, D., Sieber, A.: Performance demonstration for humanitarian demining Q – test and evaluation of mine searching equipment in detecting mines. *Mater. Test.* 45, 11, 504–512 (2003)

176. Muller, D., Uhelmann, S.: Open-cast mine optimization: maintenance. *World Min. Surf. Underground* 59, 2, 106–109 (2007)

177. Nollet, G., Prince, D.: Rotating equipment reliability for surface operation. Part II: Oil analysis in a mine. *CIM Bull.* 96, 1067, 82–86 (2003)
178. Nollet, G., Prince, D.: Rotating equipment reliability for surface operation. Part I: Vibration analysis for rotating equipment. *CIM Bull.* 95, 1065, 72–77 (2002)
179. Nussey, C.: The use of systematic reliability assessment techniques for the evaluation and design of mining electronic systems. *J. Occupat. Accidents* 7, 1–16 (April 1985)
180. Pal, B.K., Maiti, J.: Lean maintenance – concept, procedure, and usefulness. *J. Mines Metals Fuels* 52, 11, 318–321 (2006)
181. Paraszczak, J.: Failure and downtime reporting – a key to improve mine equipment performance. *CIM Bull.* 93, 1044, 73–77 (2000)
182. Paraszczak, J., Kallio, P., Honkanen, J.: Setting reliability standards for underground mining equipment. *CIM Bull.* 91, 1024, 74–79 (1998)
183. Passlow, G.: Vibration: more than a pain in backside: *Austral. Min.* 97, 10, 30 (2005)
184. Patton, P.W., Stewart, B.M., Clark, C.C.: Reducing materials handling injuries in underground mines. In: *Proc. of the 32nd Institute on Mining Health, Safety, and Research*, Salt Lake City, UT, 5–7 Aug., 1–14 (2001)
185. Petrov, N.N., Ponomarev, P.T.: Reliability of the main fan units in automatically controlled pit ventilation systems. *Sov. Min. Sci.* 6, 2, 175–179 (1970)
186. Pontt, J., *et al.*: Resonance mitigation and dynamical behavior of systems with harmonic filters for improving reliability in mining plants. In: *Proc. of the 41st IEEE Industry Applications Conference*, pp. 5–10 (2006)
187. Pontt, J., Rodriguez, J., Dixon, J.: Safety, reliability, and economics in mining systems. In: *Proc of the 41st IEEE Industry Applications Conf.*, 1–5 (2006)
188. Prasad, P.V.N., Rao, K.R.M.: Reliability models of repairable systems considering the effect of operating conditions. In: *Proc. of the Annual Reliability and Maintainability Symposium*, pp. 503–510 (2002)
189. Rahaman, M.T., Banerjee, G., Singh, V.K., Dey, N.C.: Risk assessment and prediction of roof fall of indian coal mines using seismic monitoring systems. *Indian Min. Eng. J.* April, 23–27 (2004)
190. Raju, P.G.: Longwall mining experiences at GDK No.11A mine – planned equipment salvage with particular reference to roof and side fall accidents. *J. Mines Metals Fuels* 46, 2, 92–100 (1998)
191. Ramani, R.V., Bhattacherjee, A., Pawlikowski, R.J.: Reliability, maintainability and availability analysis of longwall mining systems. In: *Proc. of the SME Annual Meeting*, pp. 1–12 (1988)
192. Randolph, R.F., Boldt, C.M.K.: Safety analysis of surface haulage accidents. In: *Proc. of the 27th Annual Institute on Mining Health, Safety, and Research*, 29–38 (1996)
193. Rugbinshtein, B.S.: Some aspects of the reliability of mining equipment. *Sov. Min. Sci.* 11, 4, 349–352 (1975)

194. Ruiz, S.E.: Reliability index for offshore pipes subjected to bending. *Struct. Safety* 2, 83–90 (1984)
195. Rushworth, A.M.: Reducing accident potential by improving the ergonomics and safety of locomotive and FSV driver cabs by retrofit. *Min. Technol.* 78, 898, 153–159 (1996)
196. Sacks, H.K., Cawley, J.C., Homce, G.T., Yenchek, M.R.: Feasibility study to reduce injuries and fatalities caused by contact of cranes, drill rigs, and haul trucks with high-tension lines. *IEEE Trans. Ind. Appl.* 37, 3, 914–919 (2001)
197. Samanta, B.M.S.K.: Reliability centred maintenance (RCM) for heavy earth-moving machinery in an open cast coal mine. *CIM Bull.* 94, 1056, 104–108 (2001)
198. Sammarco, J.J.: Addressing the safety of programmable electronic mining systems: lessons learned. In: *Proc. of the IEEE 37th Industry Applications Society Annual Meeting*, 692–698 (2003)
199. Schoolderman, A.J.: Detection performance assessment of hand-held mine detection systems in a procurement process: test set-up for MDs and MD/GPRs. In: *Proc. of the SPIE Conf. on Detection and Remediation Technologies for Mines and Minelike Targets X*, pp. 912–918 (2005)
200. Schoolderman, A.J., Roosenboom, J.H.J.: Detection performance assessment of hand-held mine detection systems in a procurement process: test set-up for MDs and MD/GPRs. *Proc. SPIE* 5794, 912–918 (2005)
201. Schroeder, C.: Reliability of main fans in coal mines. *Glueckauf Translat.* 122, 20, 367–370 (1986)
202. Scott, A.: Killing off errors. *Mine Quarry* 24, 9, 14–18 (1995)
203. Seal, A.B., Bise, C.J.: Case study using task-based, noise-exposure assessment methods to evaluate miner noise hazard. *Min. Eng.* 54, 11, 44–48 (2002)
204. Shpiganovich, A.N., Maslovskaya, T.N.: Effect of the reliability of electrical equipment on the parameters of sectional bulk transport and the performance of mining machinery. *Sov. Min. Sci.* 21, 6, 503–506 (1982)
205. Simpson, G.C.: Promoting Safety Improvements via Potential Human Error Audits, *Min. Eng.* 154, 395, 38–42 (1994)
206. Singh, B.P., Tiwari, S.K.: Application of reliability and availability to underground mine transport. *Indian J. Mines Metals Fuels* Jan–Feb., 23–30 (1984)
207. Smith, A.: Exploding the myth of condition monitoring and preventative maintenance. *Min. Technol.* 75, 864, 105–111 (1993)
208. Soldatov, A.N.: Reliability of blasting equipment. *Sov. Min. Sci.* 5, 5, 518–523 (1969)
209. Sottile, J.J., Holloway, L.E.: Overview of fault monitoring and diagnosis in mining equipment. *IEEE Trans. Ind. Appl.* 30, 5, 1326–1332 (1994)
210. Sottile, J.J., Kohler, J.L.: On-line method to detect incipient failure of turn insulation in random-wound motors. *IEEE Trans. Energy Convers.* 8, 4, 762–768 (1993)
211. Spanton, H.M., Blelloch, J.D.: Reliability of machines and machine systems. *Colliery Guard. Annu. Rev.* August, 495–503 (1977)

212. Steele, J.P.H., Archuleta, M., Brown, G.D., Drouillard, T., Schlief, D., Seifert, T.D.: Intelligent on-line monitoring of machine health for robots in critical environments. In: *Proc. of the 2nd Speciality Conference on Robotics for Challenging Environments*, pp. 262–275 (1996)
213. Stewart, B., Iverson, S., Beus, M.: Safety considerations for transport of ore and waste in underground ore passes. *Min. Eng.* 51, 3, 53–60 (1999)
214. Tait, R.B., Emsile, C.: The use of fracture mechanics in failure analysis in the offshore diamond mining industry. *Eng. Failure Anal.* 12, 6, 893–905 (2005)
215. Tan, Z., Cai, M., Yue, Z., Tham, L.G., Lee, C.F.: Application and reliability analysis of DPM system in site investigation of HK weathered granite. *J. Univers. Sci. Technol. Beijing Mineral Metallurgy Mater.* 12, 6, 481–488 (2005)
216. Thompson, R.J., Visser, A.T.: Mine haul road maintenance management systems. *J. So. Afr. Inst. Min. Metallurgy* June, 303–312 (2003)
217. Tien, J.C.: Health and safety challenges for China's mining industry. *Min. Eng.* 57, 4, 15–23 (2005)
218. Tomlingson, P.D.: Conducting mobile equipment maintenance. *Coal* 99, 11, 40–42 (1994)
219. Tomlingson, P.D.: Achieving world class maintenance status. *Min. Eng.* 58, 11, 30–33 (2006)
220. Tregelles, P.G., Worthington, B.: Reliability assessment as an aid to the development of mining equipment. *Min. Eng.* 142, 251, 95–102 (1982)
221. Turcic, P.M.: Health and safety implications of the use of diesel-powered equipment in underground coal mines. In: *Proc. of the 3rd Mine Ventilation Symposium*, p. 390 (1987)
222. Ursenbach, A., Wang, Q., Rao, M., Coward, J., Lamb, D.K.: Intelligent maintenance support system for mining truck conditioning monitoring and troubleshooting. *Int. J. Surf. Min. Reclamat. Environ.* 8, 2, 73–81 (1994)
223. Vagenas, N., Kazakidis, V., Scoble, M., Espley, S.: Applying a maintenance methodology for excavation reliability. *Int. J. Surf. Min. Reclamat. Environ.* 17, 1, 4–19 (2003)
224. Vagenas, N., Nuziale, T.: Genetic algorithms for reliability assessment of mining equipment. *J. Qual. Mainten. Eng.* 7, 4, 302–311 (2001)
225. Vagenas, N., Runciman, N., Clement, S.R.: Methodology for maintenance analysis of mining equipment. *Int. J. Surf. Min. Reclamat. Environ.* 11, 1, 33–40 (1997)
226. Van Wijk, J., Latilla, J., Wevell, E., Neal, D.: Development of a risk rating system for use in underground coal mining. In: *Proc. of the 21st Int. Conf. on Ground Control in Mining*, pp. 310–313 (2002)
227. Vanmarcke, E.H., Angelides, D.: Risk assessment for offshore structures: a review. *J. Struct. Eng.* 109, 2, 555–571 (1983)
228. Vayenas, N., Yuriy, G.: GenRel – a computer model for prediction of mining equipment failures based on genetic algorithms (GAs). *Int. J. Surf. Min. Reclamat. Environ.* 19, 1, 3–11 (2005)

229. Vayenas, N., Yuriy, G.: Using GenRel for reliability reliability assessment of mining equipment. *J. Qual. Mainten. Eng.* 13, 1, 67–74 (2007)
230. Veskov, M.I.: Estimating the reliability of new methods of coal cutting. *Sov. Min. Sci.* 3, May–June, 196–199 (1965)
231. Vinnem, J.E.: Quantitative risk analysis in the design of offshore installations. *Reliabil. Eng.* 6, 1–12 (1983)
232. Walker, A.J.: Engineering reliability into AFCS. *Min. Technol.* 64, 736, 88–91 (1982)
233. Wang, Y., Lu, G.: Analysis and definition of devices' availability and operation ratio in coal-mining systems. In: *Proc. of the Mining Science and Safety Technology Conference*, pp. 83–89 (2002)
234. Wei, F., ZhangLi, B., Yu, G.: ON microcomputer monitoring system for examining reliability of fully mechanised mining equipment. In: *Proc. of the 19th Int. Symp. on Application of Computers and Operation Research in Minerals*, pp. 848–854 (1986)
235. Whyatt, J., Miller, S., Dwyer, J.G.: NIOSH computer programs for bench crest failure analysis in fractured rock. In: *Proc. of the 32nd Int. Conf. on the Application of Computers and Operations Research in the Mineral Industry*, pp. 439–446 (2005)
236. Wilkie, J.F.: Care and reliability of rotating mining equipment. *Min. Technol.* 56, 639, 18–19, 22–23 (1974)
237. Wilson, E.B.: Reliability engineering and replacement theory – aids to maintenance decisions. *Can. Inst. Min. Bull.* 463–471 (1970)
238. Winkelmann, C.A., Craig, A., Kissel, A.M., Nutter, R.S.: Microprocessor self-testing for mine monitor systems. *IEEE Trans. Ind. Appl.* 1A–21, 1, 158–161 (1985)
239. Wirsching, P.H.: Fatigue reliability for offshore structures. *J. Struct. Eng.* 110, 10, 2340–2356 (1984)
240. Wirsching, P.H., Stahl, B., Nolte, K.G.: Probabilistic fatigue design for ocean structures. *J. Struct. Div. Am. Soc. Civil Eng.* 103, ST10, 2049–2062 (1977)
241. Wood, J.: Explosive equipment. *Electr. World* July, 22–23 (2001)
242. Worthington, B.: Reliability assessment of mining electronic systems. *Min. Eng.* 144, 277, 233–236 (1984)
243. Yamshchikov, V.S., Blok, A.V.: Information parameters and estimates of the reliability of geomonitoring of the state of the rocks. *Sov. Min. Sci.* 12, 1, 114–117 (1976)
244. Zel'vyanskii, A. Sh., Kaufman, L.L., Miroshnikov, S.I.: A method of sequential optimization of the distribution of reliability indices among the components of the transport network of a mine. *Sov. Min. Sci.* 5, 532–536 (1976)
245. Zeng, W.: Exploration for human factors in the design of coal-mine safety and rescue devices. In: *Proc. of the 7th Int. Conf. on Computer-Aided Industrial Design and Conceptual Design*, pp. 3–4 (2006)
246. Zhang, L., Luo, S., Chen, W.: The reliability analysis of the NONEL Detonating system in open-pit. In: *Proc. of the 3rd Int. Symp. on Mine Planning*, pp. 727–733 (1994)

247. Zhang, R.X., *et al.*: Surface mine system simulation and safety risk management. *J. China Univ. Min. Technol.* 16, 4, 413–415 (2006)
248. Zhitkov, E.F.: Use of the theory of random functions to assess the reliability of roof control by smooth caving onto yielding chain pillars. *Sov. Min. Sci.* 8, 1, 63–65 (1972)

Author Biography

Dr. B.S. Dhillon is a professor of engineering management in the Department of Mechanical Engineering at the University of Ottawa. He has served as chairman/director of the Mechanical Engineering Department/Engineering Management Program for over 10 years at the same institution. He has published over 335 (i.e., 195 journal + 140 conference proceedings) articles on reliability, maintainability, safety, engineering management, etc. He is or has been on the editorial boards of nine international scientific journals. In addition, Dr. Dhillon has written 33 books on various aspects of reliability, maintainability, design, safety, quality, and engineering management published by Wiley (1981), Van Nostrand (1982), Butterworth (1983), Marcel Dekker (1984), Pergamon (1986), etc. His books are being used in over 75 countries, and many have been translated into languages such as German, Russian, and Chinese. In 1987, he served as general chairman of two international conferences on reliability and quality control held in Los Angeles and Paris.

Professor Dhillon has served as a consultant to various organizations and bodies and has many years of experience in the industrial sector. At the University of Ottawa, he has been teaching reliability, quality, engineering management, design, and related areas for over 28 years and has also lectured in over 50 countries, including keynote addresses at various international scientific conferences held in North America, Europe, Asia, and Africa. In March 2004, Dr. Dhillon was a distinguished speaker at the Conference/Workshop on Surgical Errors (sponsored by the White House Health and Safety Committee and the Pentagon) held on Capitol Hill (One Constitution Avenue, Washington, D.C.).

Professor Dhillon attended the University of Wales, where he received a B.S. in electrical and electronic engineering and an M.S. in mechanical engineering. He received a Ph.D. in industrial engineering from the University of Windsor.

Index

H

I

J

L

M

N

O

P

Q

R

S

T

U

V

W